The Doomsday Lobby

James T. Bennett

The Doomsday Lobby

Hype and Panic from Sputniks, Martians,
and Marauding Meteors

 Springer

James T. Bennett
Department of Economics
George Mason University
Fairfax, VA
USA
jbennett@gmu.edu

ISBN 978-1-4419-6684-1   e-ISBN 978-1-4419-6685-8
DOI 10.1007/978-1-4419-6685-8
Springer New York Dordrecht Heidelberg London

Library of Congress Control Number: 2010936072

Printed on acid-free paper

Springer is part of Springer Science+Business Media (www.springer.com)

Acknowledgments

I am grateful to many for their assistance with and support of the research and editing of this book. The research would not have been possible without the generous financial support of the Sunmark Foundation and help from the Locke Institute. Research assistance was provided by Steven M. Davis. I also owe profuse thanks to my editor, Bill Kauffman, for I am indebted to him for significant contributions to this study.

Contents

Chapter 1

Introduction and Overview

Federal patronage of science was never contemplated by the men who wrote the U.S. Constitution. The Founders did seek, in Article 1, Section 8, to "promote the Progress of Science and useful Art" by granting inventors patent rights, but direct subvention to scientists and scientific organizations was not considered within the realm of the central government.

Yet today, Washington has become the primary patron of American science. As in so many other areas of life and policy, the Founders would be astonished by this development. As historian Walter A. MacDougall writes, "the financial, military, and organizational demands of 'big science' tended to reinforce the national state as the most efficient *agent* of technical change."[1] Big Science demanded Big Government — and vice versa. This transformation from a science that relied on private patronage to a science that is in many ways a creature of the federal government is partly explainable in terms of the general growth of state power in all areas, but it is also the result of media-savvy campaigns by advocates of Big Science to convince legislators and taxpayers that a vast expansion of government science is necessary to meet emergencies or perform herculean tasks of dubious usefulness. This book examines episodes during which those who favor a significant role for the federal government in science have used — and are still using — crises, whether real or hyped, to further their goals.

Chapter 2 develops the history of the relationship between science and government in America, with an emphasis on the flourishing of certain fields — particularly astronomy — due to private patronage. For instance, the golden age of American telescope making, from the late nineteenth through the

J.T. Bennett, *The Doomsday Lobby: Hype and Panic from Sputniks, Martians, and Marauding Meteors*, DOI 10.1007/978-1-4419-6685-8_1,
© Springer Science+Business Media, LLC 2010

early-mid twentieth centuries, when the great observatories were built and equipped, was the product almost exclusively of the private sector, and of the often reviled "robber barons" who sought immortality — or at least respectability — by attaching their names to instruments and observatories that looked deep into the sky.

War has always been good at welding the link between Big Government and Big Science. Chapter 3 studies the period when the peacetime linkage between the federal government and science was well and truly consummated: the five years between the release of *Science — The Endless Frontier* (1945), the report to President Roosevelt by Vannevar Bush, dean of politicized scientists, and the resultant creation in 1950 of the National Science Foundation, which established in the bureaucracy an official partnership between Big Science and Big Government. This was a signal achievement of the Cold War and the communist scare. Special attention is given to eloquent upholders of laissez-faire principles, such as Frank Jewett, president of Bell Telephone and the National Academy of Sciences, who spoke a language of voluntarism in an age of coercion.

Chapter 4 examines the fallout from *Sputnik*, the Russian satellites first launched in 1957. The U.S. education establishment (assisted by the building trades of the AFL-CIO) exploited the media-fanned panic over *Sputnik* to persuade Congress to pass legislation providing for unprecedented federal involvement in what had formerly been an almost purely local and state concern: the education of children. *Sputnik* forever altered the terms of the debate over federal aid to education; indeed, it can be said to have almost ended the debate, as the traditional view of education as a province of local governments has since been muted virtually into silence.

Chapter 5 reviews the more recent attempt to galvanize the federal government in support of a fantastically expensive exhibition of Big Science: a manned mission to Mars, a persistent enthusiasm of recent American presidents in search of a crowning achievement. The Mars boosters have gained traction not so much by exaggerating threats — though some argue that we must begin to plan for the evacuation of our planet due to eventual environmental catastrophe — as they have by proclaiming it the duty of government-subsidized science to perform an interplanetary act of "national greatness" at a staggering cost to the taxpayer.

Chapter 6 discusses the often-colorful ways that media-savvy seekers of federal funds are currently using the science-fiction-like threat of killer comets and asteroids — fanciful doomsday scenarios of DEATH FROM THE SKIES! — to secure funding for favored astronomy projects. These Chicken Littles have taken a possibility of exceeding remoteness and, with diligence and a talent for scaring the wits out of laymen, magnified it to such an extent that they are carving out untouchable turf in battles over the allocation of federal science research funds. After all, their mission — preserving life on earth — trumps all other causes.

The final chapter distills the essential arguments developed in earlier chapters to show that, like oil and water, politics and science do not mix well. When government plays the central funding role in science, issues are imbued with a political correctness that impedes objective, unbiased inquiry, which is the very essence of science. Although this study focuses on astronomy, the recent controversy roiling around the issue of global warming ("Climategate") and the efforts of its proponents to ensure that the data support their claims is an excellent example of how government funding can corrupt scientific inquiry.

We should take the predictions and warnings to follow, which range from cautions that the United States will be vanquished if it does not establish science bureaucracies to alarmist tales of killer asteroids wiping out life on Earth if astronomers do not get their share of federal dollars, with more than the proverbial grain of salt. For fiction in the service of science is an old story in the annals of government-building. My colleague at George Mason University, Walter Williams, has compiled predictions made within the last half century that have not exactly come true. For instance:

> "The threat of a new ice age must now stand alongside nuclear war as a likely source of wholesale death and misery for mankind." — Nigel Calder, *New Scientist*, 1969

> "The cooling since 1940 has been large enough and consistent enough that it will not soon be reversed" — C C Wallen, World Meteorological Association, 1969

> "[I]n the 1970s ... hundreds of millions of people are going to starve to death." — Paul Ehrlich, 1968[2]

These three predictions demonstrated that their sayers were far from soothsayers. It would be hard to be more wrong than Calder, Wallen, and Ehrlich were — not that it harmed their popular reputations all that much.

Similarly, and despite the elaborate scare stories of the prophets of Big Science, the United States never did fall into a scientific Dark Age, or get overrun by communist hordes, or get taken to school by superior Soviet scientists, or lose the spirit of innovation due to our failure to send an astronaut to Mars, or get wiped out à la the dinosaurs. But the veracity of the threat-makers doesn't really matter. All that counts is that enough politicians believe them. All that counts is that the pipeline of federal dollars flows. And so far, that flood of dollars shows no sign of shutting off.

Notes

1. Walter A. McDougall, "Technocracy and Statecraft in the Space Age — Toward the History of a Saltation," *American Historical Review*, Vol. 87, No. 4 (October 1982), p. 1022.
2. Walter Williams, "Into the Wild Green Yonder," *Washington Times*, May 13, 2008.

Chapter 2

American Science before the Bomb

The patronage of science by the new national government was barely even an afterthought in the Constitutional Convention of 1787. On September 14, in the final week of the conclave, James Madison of Virginia and Charles Pinckney of South Carolina moved to give the Congress the power "to establish an University."[1] There appears to have been very little debate on the matter — or perhaps Madison, who took the most detailed notes, suffered from carpal tunnel syndrome — and the motion was defeated, with four states (Pennsylvania, Virginia, North Carolina, and South Carolina) in favor, six (New Hampshire, Massachusetts, New Jersey, Delaware, Maryland, and Georgia) opposed, and Connecticut divided. (The majority of New York's delegation had left town, and Rhode Island never sent anyone to Philadelphia.)

Thus, the Framers of the Constitution rejected the only effort to include in that document a subvention, even if indirect, of science and scientific research. The Constitution did "promote the Progress of Science and useful Art" by establishing the power to grant inventors and authors an "exclusive Right" to their creations — a patent, in other words. In this way, it gave inventors and men of science free rein to follow the muse and profit from whatever might flower from their genius.

The early sessions of Congress saw sporadic attempts by advocates of a more energetic national state to bring scientific research within the purview of the new government. Yet in the very first Congress, Maryland Rep. Michael Jenifer Stone, referring to a brief discussion of establishing a national university, to which there was overwhelming opposition, said, "We

J.T. Bennett, *The Doomsday Lobby: Hype and Panic from Sputniks, Martians, and Marauding Meteors*, DOI 10.1007/978-1-4419-6685-8_2,
© Springer Science+Business Media, LLC 2010

have already done as much as we can with propriety; we have encouraged learning, by giving to authors an exclusive privilege of vending their works; this is going as far as we have power to go by the Constitution."[2]

"As far as we have the power to go by the Constitution": it's been a long time since that kind of language was common in the Congress.

Seekers of federal backing for scientific projects had little luck when importuning Congress. For instance, a February 3, 1796, petition from Frederick Guyer seeking "pecuniary aid from Government to enable him to prosecute his researches for the discovery of the longitude by lunar observations" was rejected out of hand by the relevant committee "as unconstitutional."

Rep. John Page, a Jeffersonian Republican of Virginia, expressed the hope that by flatly rejecting Guyer's entreaty the Congress would lay down, once and for all, the correct constitutional position and "save Congress from similar applications."[3] No such luck, Rep. Page. Eventually, Congress would be *inviting* such applications. It would solicit the solicitors, and treat them solicitously.

This is not to say that the federal government in its early years adhered to a scrupulously hands-off policy regarding science.

The Swiss-born surveyor F.R. Hassler, a professor of mathematics at West Point, undertook the first United States Coast Survey in 1807 at the behest of that sometime strict constructionist, Thomas Jefferson. (As a strict constructionist of the Constitution, Jefferson believed that a constitutional amendment would be necessary to permit the feds to subsidize even a national museum in the federal city.) Early federal assistance to science was somewhat indirect, often in the form of aiding explorers. For instance, Lewis and Clark were said to further commerce with the West, thereby justifying federal provisioning of their journey.

Yet even the coastal survey and meteorological studies were viewed skeptically by members of Congress inclined to a strict constructionist view of the Constitution. Open the till to pay for even such reasonable activities, they charged, and you will never be able to close the till again. Moreover, the sectionalism that marked 19th-century American politics was at play, as representatives of inland states or districts contended, with justification, that a coastal survey or other aids to navigation were artificial boons to eastern seaboard commerce.

As Americans had become politically independent of Europe, so did they seek not scientific *independence*, for science knows no national boundaries, but scientific *distinction*. It was to be achieved in ways other than through royal patronage or national subsidy. There were no kings to bow and scrape to and from whom to beg money. There wasn't even a National Science Foundation.

Private organizations dedicated to the exchange and diffusion of knowledge among scientists sprang up naturally — for example, the American Academy of Arts and Sciences in Boston, the American Philosophical Society in Philadelphia, and the American Academy for the Advancement of Science (founded as the Association of American Geologists and Naturalists) in Philadelphia.

The writing, to some, was on the wall, if not the appropriations bill. As British chemist Joseph Priestley said to the American Philosophical Society: "I am confident...from what I have already seen of the spirit of this country, that it will soon appear that Republican governments, in which every obstruction is removed to the exertion of all kinds of talent, will be far more favorable to science, and the arts, than any monarchical government has ever been."[4]

The local societies of amateur scientists, which sponsored lectures and sometimes even publications, which brought together men and women of kindred interests, did much to spread the spirit of scientific inquiry throughout the land. "In 1846, New England had fourteen of the nation's thirty-two existing scientific societies," with most of the rest (eleven) in the Middle Atlantic states of New York, New Jersey, and Pennsylvania. Science was finding its level, with no help from Washington. In fact, most men of science had no intention of seeking alms from the government. As Robert V. Bruce writes in *The Launching of Modern American Science: 1846–1876* (1987), "Scientists tended to be self-sufficient, self-controlled, independent-minded, assertive, even domineering."[5] They were not the sort to whimper and purr for state favors with a self-diminishing docility.

Yet in Washington the first stirrings of the eventual, if uneasy, merger of government and science could be heard in the debate over the Naval Observatory.

Light-Houses of the Skies

The first great battle over the role of the federal government in science was fought more on celestial than on terrestrial grounds. At issue: whether or not the government should build a national observatory to scan the skies.

For the first several decades of the American republic, the consensus was that such a project went well beyond what was constitutionally permissible. In a pleasing contrast to later debates, its advocates argued on grounds of merit. They had not learned — or maybe they just disdained — the use of threats, scare tactics, and boogeymen. No one stooped to argue that if the federal government did not fund an observatory, Martians would sail from their red planet, heavily armed and hostile in intent, to colonize ours.

Among coastal survey or surveyor F.R. Hassler's recommendations was the establishment of astronomical observatories in Maine and southern Louisiana — as far north and as far south as conditions would permit. Their purposes would be largely navigational — that is, they would refine the measurements of time and longitude and thereby assist the ships at sea. Hassler envisioned these as "permanent scientific establishments," but he tipped his hand when speculating on the possibility of placing one of the observatories not at the nation's extremities but in its capital city. "[V]arious considerations," he said, "might occasion and favor the desire of placing one of these observatories in the city of Washington, as observatories are placed in the principal capitals of Europe, as a national object, a scientific ornament, and a means of nourishing an interest for science in general."[6] Hassler had surveyed the political as well as the coastal terrain. It makes sense to butter up the funders.

The Jeffersonian Republicans who were then the dominant party did not respond well to the idea of the national government paying for "ornaments" to its greatness, so even though Hassler picked a spot for the observatory near the Capitol and although President Madison gave the plan his approval in 1816, it died aborning. A subsequent reauthorization of the coastal survey in 1832 stated that nothing in the act should be interpreted "to authorize the construction or maintenance of a permanent astronomical observatory."

As historian Dr. Charles O. Paullin explained in a paper read to the Columbia Historical Society of Washington, DC in 1921, Hassler was not the only dreamer of supernal dreams. In 1809, William Lambert played to patriotic sentiments in requesting that Congress take steps to create a prime meridian in the United States, rather than relying on the British meridian at Greenwich. This was a canny strategy, as it "seemed to involve a declaration of astronomical independence from Great Britain." As Paullin writes, "A native republican meridian was to be substituted for an alien monarchical one."

Given that the United States was in the midst of an extended conflict with Great Britain, which had begun with an embargo in 1807 and would culminate in the War of 1812, Lambert chose a fine time to exploit anti-British sentiment in the service of government subsidy. James Monroe, President Madison's Secretary of State, declared that "An observatory [to establish the meridian] would be of essential utility."[7] A House committee agreed, reporting out a bill in January 1813 to do just that, but the observatory, which was to be directed by the grandly named National Astronomer and built in Washington — not as an ornament but as a patriotic necessity! — was lost in the chaos of the war.

The idea of a national observatory got its strongest push from John Quincy Adams, sixth president of the United States, and an avid amateur astronomer for whom "the founding of a national observatory became one of the cherished projects of his later life."[8]

President Adams, in his inaugural address of March 4, 1825, had pledged fealty to the U.S. Constitution, that "revered instrument" which enumerated and delimited the powers of his office.[9] One man's fealty is another man's license, however, for in his first annual message to the Congress on December 6, 1825, Adams issued a rousing call for a national observatory and the creation of a position similar to the aforementioned National Astronomer. Said Adams:

> It is with no feeling of pride as an American, that the remark may be made that on the comparatively small territorial surface of Europe there are existing upward of 130 of these light-houses of the skies, while throughout the whole American hemisphere there is not one. If we reflect for a moment upon the discoveries which in the last four centuries have been made in the physical constitution of the universe by means of these buildings and of observers stationed in them, shall we doubt of their usefulness to every nation? And while scarcely a year passes over our heads without bringing some new astronomical discovery to light, which we must fain receive at second-hand from Europe, are we not cutting ourselves off from the means of returning light for light, while we have neither observatory nor observer upon one half of the globe, and the earth resolves in perpetual darkness to our unsearching eyes?[10]

Adams foreshadows themes that will be present in later shakedowns of the public treasury for scientific research. First of all, Europe is doing it! This argument can cut both ways. Sometimes the speaker will appeal to American exceptionalism by urging his countrymen not to follow the example of the decadent old continent; at other times, as in Adams's address, the appeal is made to national pride to catch up to Europe, to show those haughty royalists that republican and democratic Americans are just as capable as the toffs and serfs of the Old World.

A second theme limned by Adams is that of missed opportunity, which often — though not in this case — is tinged with the hint of potential catastrophe. The mysteries of the universe are unrevealed unless and until the taxpayers of the states shell out for a national observatory. Our "unsearching eyes" condemn us to "perpetual darkness," and we all know that dangers lurk in the dark. The problem is that the lethargic and incurious Americans are not at the "curse the darkness/light a candle" stage. They actually seem *comfortable* in the darkness, or at least they are unwilling to take the administrative steps necessary to erect a great lighthouse of the skies whose beacon, coincidentally, will shine from and upon the capital city of the new country.

This was the centerpiece of President Adams's ambitious plan to place the federal government in a lead role in the cultivation and promotion of scientific knowledge. Other new activities the president urged upon the national government included a national university, a naval academy to complement the military academy at West Point, and a wide array of "internal improvements," i.e., the federal sponsorship of roads, bridges, and canals.

The Adams proposal received a favorable report from a House committee but no action was taken in the full House. He was mocked for his poetic phrase "light-houses of the skies." As President Adams learned, never use poetry in a political speech when a prosy platitude will suffice.

The odd thing was that Adams had, just two years earlier, indicated an alternative path to the comprehension of the heavens when he "offered to give a thousand dollars towards the establishment of an observatory at Harvard University."[11] Harvard would, in time, build from private monies an observatory whose work was a brilliant blazing sun in comparison to the later government observatory's unprepossessing ort of dust. But when tax revenue is at your fingertips, spending it seems essential.

Attorney General William Wirt and Secretary of State Henry Clay both had constitutional objections to Adams's grand vision of a federal government that endows a national university and observatories. Moreover, voters were unenthusiastic for fear that this portended, says author A. Hunter Dupree, "consolidated government, monarchy, and tyranny."

"The bold attempt to assert the constitutionality of central scientific institutions and to tie them to a vigorous exercise of power by the central government had failed resoundingly," continues Dupree in *Science in the Federal Government: A History of Policies and Activities to 1940* (1957).[12] Nevertheless, despite the lack of legislative support, a naval observatory eventually came into being as an adjunct to the navy, which pleaded for help in timekeeping and weather forecasting.

Indeed, the city of Washington got its observatory in what would prove to be the time-tested method: via the military. Specifically, the navy, whose Lieutenant James M. Gillis put in herculean work nights in the late 1830s to record all manner of celestial phenomena: the movements of the planets, the stars, the moon, and transits across the surfaces thereof by other bodies. His observations and those of his assistants, which were conducted in a small building on Capitol Hill, so impressed the Navy that its secretary requested of President Tyler and Congress in 1841 the appropriation of funds to build an observatory — though the word used was "depot," which sounded much more functional and utilitarian — for these vital observations.

A bill to do that passed in the last hour of the 1841–42 session, and so on August 31, 1842, the construction of the Depot of Charts and Instruments was mandated by law. This was an observatory in all but name, and in fact when the clouds cleared the depot was renamed the United States Naval Observatory and Hydrographical Office. The latter office was split off, and so, after decades of dreaming, John Quincy Adams had his government-subsidized light-house of the sky: the United States Naval Observatory.

Adams said, somewhat puckishly, that he was "delighted that an astronomical observatory — not perhaps so great as it should have been — had been smuggled into the number of institutions of the country, under the mask of a small depot for charts."[13]

Dissimulation, dissembling, deception: no matter how it had been accomplished, the federal government had its observatory.

Now, at the same time that Adams and company were lobbying so heavily for a taxpayer-subsidized observatory, private observatories were springing up at Western Reserve College (1838), Harvard College (1839), Philadelphia High School (1840), and most startlingly, in Cincinnati. By 1882, the United States boasted 144 astronomical observatories, almost all of them nongovernmental.

The Naval Observatory would enjoy its salad days under the directorship of Simon Newcomb (1835–1909), a Canadian-born publicity genius who wrote popular books on astronomy and assembled a "large staff of assistants" courtesy of the federal government.[14] But the real action would be far outside the walls of government science, and it would be the work of the dynamic pairing of energetic astronomers and philanthropic capitalists. But more on that in a bit.

Smithson's Gift

British chemist James Smithson's bequest upon his death in 1829 that after his third nephew died he would leave "to the United States of America, to found at Washington, under the name of the Smithsonian Institution, an Establishment for the increase & diffusion of Knowledge among men," caused a great debate, in part over whether or not the government could even accept such a gift.[15] It's not that Americans were disposed to look an English gift horse in the mouth, it's just that they wondered if Smithson's legacy might also include the key to Pandora's box. In any event, Smithson's bequest was accepted in the Senate by a vote of 31–7 while the House voted 115–8 against a motion to return Smithson's money.

John Quincy Adams, as usual, was at the center of the debate. Adams, who served in the U.S. House of Representatives after his single term in the White House, was chairman of the committee to which the presidential message announcing Smithson's bequest was referred. He thought that the money might be used for the "establishment of an Astronomical Observatory, with a salary for an astronomer and assistant, for nightly observations and periodical publications; annual courses of lectures upon the natural, moral, and political sciences."[16] Said Congressman Adams, "There is no richer field of science opened to the exploration of man…than astronomical observation."[17] He was nothing if not single-mindedly stubborn. Congressman Adams also knew what he did not want: "Above all no jobbing, no sinecure, no monkish stalls for lazy idlers." What he would have thought of the National Endowment for the Humanities we leave up to the imagination of the reader.

For the next five years, Adams did what he could to steer some of Smithson's money toward the construction of a light-house of the sky. He wrote letters, he testified, he introduced bills, he even invited the advice of George Airy, the Astronomer Royal of Great Britain and director of the observatory at Greenwich, and a snooty character sure to repel those Americans who had not forgotten that their country had fought two wars against England over the last 60 years. In 1840, Adams, in "a display of much learning and some rhetoric," instructed the House in the history of astronomy from Genesis until 1839.[18] No one ever said the old man was not pedantic.

Adams was still in the House when in 1846 the Smithson monies were distributed. Not a penny went to build an astronomical observatory.

It might do us good to listen to the far-off voices of those who did not want Mr. Smithson's money. They will strike our ears as odd, but they were very much part of the constitutional chorus.

Rep. John Smith Chipman (D-MI) asked in April 1846, "How did it happen that this Government accepted such a boon from a foreigner — an Englishman, too. He looked upon it as a stain on the history of the country, as an insult to the American nation. He wished this Government to wash its hands of all such eleemosynary dealings. There was a native stock in this country, intellectual and physical, that needed no foreign aid, and he trusted in God it would not condescend to receive any. He objected to the bill, because, clearly and in terms, it established a corporation. He appealed to his political friends, after all their opposition, after all their arguments, after all their efforts to put down a United States Bank on the ground of its unconstitutionality, whether — tickled, amused, their pride touched by the great advantages of dispelling the cloud of ignorance which overshadowed the American Republic — they would now belie all their principles and their professions?"

Five days later, Rep. Alexander D. Sims (D-SC) took off after Smithson. "He saw in the will of that individual what he had seen in the wills of many other men. After having griped, through their lives, every shilling that came into their hands, animated at last by some posthumous vanity, they sought to build up a name, which should live after them; and such, rather than any feeling for humanity, so much lauded, was the motive that guided them. In the present case he saw abundant evidence of this disposition in the appointment of the Government of the United States as a trustee to carry out this splended vanity." Sims also insisted that "There was no grant of power in the Constitution admitting" the "exercise" of the "administration of charities." He proposed to return the money to the British chancery. Thanks but no thanks, said Sims. Not if the cost were the Constitution.

Rep. Andrew Johnson (D-TN), the future president, "replied that he objected to the whole principle of the measure, and that he would send the money back to the source from whence it came." Coincidentally, the future President was answered by another future President, Jefferson Davis, who argued that "The Government was bound, after solemnly accepting the trust, to execute it faithfully."[19]

It took ten years for the Congress to reconcile itself to Smithson's gift, but it did. Yet in the early 19th century, while presents might be accepted, not every petitioner for federal subsidy received a check. Washington knew how to say no. It could say no even to the most fantastic among the suppliants and the dreamers. It could say no to the forerunners of our 21st century Killer Asteroid/Let's Colonize Mars crew.

Take, for instance, Captain John Cleves Symmes, Jr. and his apostle Jeremiah N. Reynolds.

Symmes, a brawler who had distinguished himself for bravery in the War of 1812, was thought by some to be a nut but he was also called "a man of decided ability, and a bold and original thinker," by P. Clark in the pages of the *Atlantic Monthly*.[20]

The duo of Symmes and Reynolds collaborated in propagating one of the wildest scientific speculations of the early American republic: to wit, that the Earth is hollow. It is composed, said Symmes, of five concentric spheres at the center of which is a hollow core. Apertures at the poles led to this center, and in fact a traveler "might pass from the outer side of the earth over the rim and down upon the inner side a great distance before becoming aware of the fact at all."[21] As evidence, Symmes pointed to reports from explorers of birds that migrate north, of reindeer that move poleward in November, of waters that seemed warmer the closer one came to the North Pole, of Indians in Baffin Bay who told explorers that their home was to the north, and that the farther south one went the colder it got.

He mailed to as many institutions of science and learning as he could afford the following circular:

Light gives light to light discover ad infinitum.

ST. Louis, MISSOURI TERRITORY, NORTH AMERICA, April 10, 1818.
I declare the earth is hollow, habitable within; containing a number of solid concentrick spheres; one within the other, and that it is open at the pole twelve or sixteen degrees. I pledge my life in support of this truth, and am ready to explore the hollow if the world will support and aid me in the undertaking.

JOHN CLEVES SYMMES OF OHIO,
Late Captain of Infantry.

N. B. I have ready for the press a treatise on the principles of Matter, wherein I show proofs of the above proposition, account for various phenomina [sic], and disclose Dr. Darwin's "Golden Secret."
My terms are the patronage of this and the new world, I dedicate to my wife and her ten children.
I select Dr. S. L. Mitchell, Sir H. Davy and Baron Alexander VonHumbolt as my protectors. I ask one hundred brave companions, well equipped to start from Siberia, in the fall season, with reindeer and sledges, on the ice of the frozen sea; I engage we find a warm and rich land, stocked with thrifty vegetables and animals, if not men, on reaching one degree northward of latitude 82; we will return in the succeeding spring.

J. C. S.[22]

One might think that such a circular would find its way into the nearest circular file of whichever office it happened to land in, and surely this was true in many instances, but Symmes's theory found its defenders. James McBride, a trustee at Miami University of Ohio, "used his able pen in arranging and elaborating Symmes' somewhat disorderly argument," as John Weld Peck wrote in a 1909 essay in *Ohio History*.[23]

Jeremiah N. Reynolds was one acolyte of Symmes's whose energy carried the hollow-earth theory to the coasts and cities. The two men, both Ohioans, undertook lecture tours that exposed crowds of upwards of a thousand people to this unorthodox theory. They proposed that the U.S. government subsidize the exploration of the poles as way of finding the center of the earth, not only for the thrill of discovery but also, as historian Robert F. Almy writes, as a way of "extending the American frontier to the ends of the earth."[24]

Symmes first asked Congress to pay for this journey to the center of the earth in 1822. None other than John Quincy Adams endorsed the Symmes request — it seems that lighthouses of the sky were not the only scientific venture Adams believed deserving of federal sponsorship. (Adams's successor, Andrew Jackson, who had a somewhat more practical knowledge of adventuring and held to a stricter construction of the Constitution, had no such enthusiasm.)

"As a lecturer," wrote Peck, Symmes "was far from a success. The arrangement of his subject was illogical, confused, and dry, and his delivery was poor." Yet so compelling was his thesis that he attracted audiences in the Ohio Valley and Kentucky, and it was a Kentucky member of Congress, Richard M. Johnson, who proposed to that body that the federal government equip Symmes and a company in the manner he had requested for a polar expedition to find the hole that leads to the middle of the earth. A motion to refer the petition to the committee on commerce — the reason given being that Americans surely would engage in commerce with any creatures found living within the earth! — attracted a surprisingly high 25 votes, though it was not enough to win the referral.

Critics mocked "Symmes' Hole," and those who wrote of it later called it an "absurd, foolish theory."[25] It is, however, the contention of Robert F. Almy that Edgar Allen Poe borrowed Symmes's notions as the basis of two of his most famous stories, "Ms. Found in a Bottle" and "A Descent into the Maelstrom."[26] After Symmes's death, Jeremiah N. Reynolds did undertake a privately financed sealing voyage toward the South Pole, but he was marooned in the Shetland Islands. Reynolds never found the opening that led to the center of the earth, but at least he proved that even such fanciful ventures can find private financing.

What the War Grew

"The Civil War blighted American science," wrote historian of science Robert V. Bruce in *The Launching of Modern American Science 1846–1876*, for while the government is always of need of advanced weaponry, outside of narrow military applications science is largely ignored for the more pressing matter of winning the war.[27] Journals were suspended or died during the war; research went undone, college classes were unattended. Money for state geological surveys dried up, or more accurately was funneled to the war effort. The Naval Observatory not only lost its director, Commander Matthew F. Maury, to the Confederate army, it was tasked with maintaining spyglasses, compasses, and the like for the Union army, which cut down greatly on the scientifically useful work it might otherwise do. The Coast Survey prospered, as defense of the coasts was essential to the Union war effort, but its labor was now singularly directed toward military application.[28] There is no time for pure science when the cannons blast and the minie balls fly.

During the Civil War, government used scientific research to assess and improve naval warfare, gunnery, and ordnance, yet among the bureaucratic fruit of the war was not a Department of Science but a Department of Agriculture, whose first commissioner was the deliciously named Isaac Newton. (He seemed to have had no apple-dropping epiphany, uneventfully overseeing the "disciplines of chemistry, botany, and entomology" as they applied to the nation's farms.[29])

Though America was born as an agrarian republic, its agrarian base arose naturally, without the sponsorship of government. As an early history of the Department of Agriculture — DOA, to use the gruesome acronym the department avoids — states, "For about sixty years subsequent to the Revolution, the general interests of agriculture were left almost entirely to individual initiative." Congress did not even bother to set up committees on agriculture until 1820 (the House) and 1825 (the Senate). In 1839 it appropriated $1,000 for "the collection of agricultural statistics, and for other agricultural purposes," and this description — the first half strictly limited, the second half ominously open-ended — sufficed for the remainder of the antebellum period, as the federal government collected data and distributed seeds and that was pretty much that.

Then war came, as President Lincoln said, and among the many changes wrought in the new relationship between Washington and the people was the 1862 creation of the Department of Agriculture, which expanded the scope of the federal government's interest in matters of the land by giving this new department the power "to acquire and to diffuse among the people of the United States useful information on subjects connected with agriculture in the most general and comprehensive sense of that word, and to procure, propagate, and distribute among the people new and valuable seeds and plants."[30]

Again, the grant power seemed both limited (what's a few seeds?) and infinite (what is "the general and comprehensive sense" of agriculture?). Also in 1862, land-grant colleges were authorized by the Morrill Land Grant

College Act, a grant of public lands to states "which may provide colleges for the benefit of agriculture and the mechanic arts."[31] Under cover of the emergency of war, a bureaucracy was growing.

The commissioner in charge of the Department of Agriculture would not have a seat at the Cabinet table until 1889, by which time the "general and comprehensive" meaning of agriculture had been interpreted to permit the federal government to investigate diseases afflicting everything from cattle to fruit trees. The department's portfolio grew so large that it was split into six sections — chemical, seed, entomological, statistical, microscopical, and botanical — and it hasn't looked back since.

The DOA was a useful precedent and harbinger of bureaucratic things to come. As historian A. Hunter Dupree noted, "Although opponents could and did invoke states' rights against federal scientific activity, the outbreak of the Civil War had ruled in favor of Alexander Hamilton's interpretation of the general-welfare clause as clearly as it presaged the triumph of Hamilton's vision of an industrial nation."[32]

The war produced the shell of what might have become a Ministry of Science, had not the American preference for private over governmental action remained strong. For from the bowels of the Union war effort came another push, this one culminating in the National Academy of Sciences, a grand title, to be sure, although the actual influence of this war creation, an advisory council of leading scientists to the government, was meager.

The bill establishing a National Academy of Sciences was approved by both houses via voice vote. Passed almost without notice under the chief sponsorship of Senator Henry Wilson (R-MA), the legislation created a 50-man academy to advise the federal government on matters scientific. President Lincoln signed the NAS bill into law on March 3, 1863, while he had more important matters on his mind.

The first NAS appointments were made for life, which was an aristocratic act almost beyond comprehension. Several of those ordained criticized sharply the elitist cast and the absurdity of the Chosen 50. Yet the first meeting of the academy, on April 22, 1863, six weeks into its existence, attracted 32 members to New York City. Charter members included such illustrious names as paleontologist Louis Agassiz; astronomer Frederick Barnard; mineralogist James Dwight Dana; chemist Wolcott Gibbs; Joseph Henry, first Secretary of the Smithsonian Institution; and West Point engineer Dennis Hart Mahan. Two-thirds of that initial class was in the fields of math and physics.

Though Senator Wilson sought a federal appropriation for the NAS, he failed to win one, and in fact, the academy would not go on the federal appropriation dole until 1941. Members paid dues rather than received stipends.

Those members were required to take the same oath of allegiance as prescribed by the U.S. Senate for senators. To criticism that scientists ought to take oaths to, say, truth or free inquiry rather than government, geologist J. Peter Lesley, one of the earliest promoters of the academy idea, said on April 23, 1863, that the oath "was only an unfortunate way of stating a higher truth, that we are the children of the government, and the Academy is the creation of the government, and owes it the oath of allegiance as its first duty."[33] Unlike such contemporaneous private associations of scholars as the American Philosophical Society, the American Academy of Arts and Sciences, and the American Association for the Advancement of Science, the NAS was beholden to the central state that gave it life.

The NAS was not exactly overburdened with work, though it did busy itself in its early years with promotion of the metric system, which some might regard as a dubious accomplishment. Besides its publicity work on behalf of meters and liters, in its first half century the NAS prepared reports — some useful, some not — on protecting the bottoms of iron ships against saltwater corrosion, observing the transits of Venus of 1874 and 1882, studying the manufacture of sugar from sorghum, analyzing the qualities of different kinds of wool to determine tariff rates, and assessing the prospects of scientific exploration of the Philippines, which the U.S. had recently acquired from Spain in the Spanish–American War.

Despite its impressive-sounding title and prestigious membership, the National Academy of Sciences lacked even a building, let alone a lab or journal. It was relatively inactive, being asked for help by the federal government just 51 times in its first 52 years. This embarrassed George Ellery Hale, whom we shall meet shortly. Hale made a critical lapse in judgement — at least from the point of view of those who prefer science to be the realm of the private sector rather than the government. In 1915, as Europe was plunged into war and agitation was growing stateside for a U.S. intervention, Hale declared that for the NAS to achieve "great results" it "must enjoy the active cooperation of the leaders of the state."[34]

This declaration ran counter to his own experience, as we shall see. Hale had found and demonstrated the generosity of the private sector, whether altruistically inclined foundations or eccentric millionaires hoping for nomenclatorial immortality. Indeed, the mediocrity of the NAS stood in stark contrast to the brilliance of what George Ellery Hale would accomplish without a dime of Washington's money.

Light-Houses of the Sky — Privately

The first American private light-house of the sky was raised at Williams College in 1838. A dozen others followed in the next 20 years. The two great observatories of the 1840s were erected by "private munificence" at Harvard and…Cincinnati.

"Upstart Cincinnati, not the older, richer cities of the Atlantic coast, built the first major American astronomical observatory," writes Howard S. Miller, and therein lies a story of pre-bureaucratic, nongovernmental science.[35]

The man who put Cincinnati's eyes on the skies was Ormsby MacKnight Mitchel, an Ohio-bred West Point grad, attorney and Professor of Mathematics, Natural Philosophy and Astronomy at Cincinnati College. A fluent lecturer, Mitchel rallied the citizens of his city to the cause of an observatory. He deployed his patriotism expertly. Mitchel was, he said, "determined to show the autocrat of all the Russias that an obscure individual in this wilderness city in a republican country can raise here more money by voluntary gift in behalf of science than his majesty can raise in the same way throughout his whole dominions."[36] And he did, equipping his new observatory with one of the best refracting telescopes anywhere. (Alas, his college, though not the telescope, soon burned to the ground!)

Mitchel's was, as Stephen Goldfarb wrote in "A History of the Cincinnati Observatory, 1842–1872," "an exemplary story of science and democracy" which was founded "in a curious blend of localism, nationalism, and… civic pride." He might have added "private initiative," for "The monies for the telescope and construction of the observatory were raised by public subscription, making this the first astronomical institution of its size to be built without either royal or governmental patronage."[37]

Professor Mitchel understood that private industry and enterprise were the motive forces of American achievement. And he knew something about motive forces, having served as the Little Miami Railroad's chief engineer in the months when school was out. Despite President Adams's poetic language about light-houses of sky funded by the federal government, Ormsby MacKnight Mitchel determined to raise up Cincinnati's lighthouse with contributions from the citizens of Cincinnati. After a successful series of lectures on the solar system in 1841–1842 for the earnestly titled Society for the Diffusion of Useful Knowledge, Mitchel "resolved to raise seven thousand five hundred dollars" from the sale of 300 shares of stock, which in practice were memberships in the Cincinnati Astronomical Society. This stock would not pay dividends in the form of dollars but instead access to useful knowledge, for stockholders would have "the privilege of examining these beautiful and magnificent objects through one of the finest glasses in the world."

Mitchel's audacious plan of selling stock in useful knowledge found favor among "all classes." He was not dependent upon a single capricious sugar daddy; most purchasers of the stock bought a single share at $25. Within weeks Mitchel had sold all 300 shares and more. His patrons sent him on a scouting trip to Europe, where he learned the ins and outs of using a large telescope from the Astronomer Royal and ordered the 11 and ½ inch refractor telescope from the Munich shop of Merz and Mahler. In the year that it took for Merz and Mahler to painstakingly put together this superb instrument, an observatory was constructed by local workmen on land donated by the Cincinnati winemaker and real-estate tycoon Nicholas Longworth. "Many of the materials and much of the work done on the building were donated in exchange for shares in the society," writes Stephen Goldfarb, making it a truly democratic institution. None other than John Quincy Adams spoke at the cornerstone-laying ceremony — for two hours.[38]

The Cincinnati Observatory opened in early 1845. For the first year-plus of its existence, the telescope was available to members from 3 p.m. to 10 p.m. every day but Sunday and Monday, though this was cut back to public viewings on Thursday, Friday, and Saturday in late 1846 in order to facilitate the research projects of the observatory. Still, it had achieved a balance between the diffusion of knowledge to interested Cincinnatians and the undertaking of original research in astronomical science – a balance that lasted almost ten years, till in 1854 the observatory devoted itself purely to research.

Mitchel moved on to the Dudley Observatory in Albany in 1860; the Cincinnati Observatory came under the auspices of the University of Cincinnati and was moved to a more advantageous spot. Alas, Ormsby Mitchel — Brigadier General Mitchel to his troops, who called him "Old Stars" — died of yellow fever in South Carolina in October 1862. Old Stars had shown that community pride was sufficient to build a grand observatory in a western town.

Bostonward, Benjamin Pierce, professor of mathematics and astronomy at Harvard, also appealed to civic pride: "Massachusetts is not wont to lag behind. I trust it is only necessary that the public should know the want to relieve it." Relieve it they did, and Harvard soon had its observatory, though not before the frontier city on the Ohio River. Cincinnati raised its light-house of the skies by appealing to the local pride and scientific curiosity of the man on the street; Harvard took a different tack. Howard S. Miller says of the men who rose to the occasion in Cambridge: "These patrician philanthropists, while eager to peer through a telescope, were also willing to support men of science in their tedious, abstruse, often unspectacular investigations."[39] Taking note of the flowering of private science was Edward Everett, orator extraordinaire, or at least orator of extraordinary length — he was the gentleman who spoke before Lincoln at Gettysburg, going on for two hours while the President's famous address lasted just two minutes, though the world has little noted or long remembered the longer-winded of the speechifiers. Everett spoke on "The Uses of Astronomy" at the dedication of the Dudley Observatory in Albany, which had been largely endowed by the generosity, to the tune of $75,000, of Mrs. Blandina Dudley. Everett, true to his loquacious reputation, spoke for what must have seemed a Saturnian year at Albany, despite his de rigueur false modesty in saying that the keynote address "ought to have been elsewhere assigned," to some eminent scientist. Everett could play second fiddle to no one. Although he was very much a Whig who believed that John Quincy Adams had been right to call for federal subvention of light-houses of the skies, he conceded that "With the exception of the observatories at Washington and West Point, little can be done, or be expected to be done, by the government of the Union or the States; but in this, as in every other department of liberal art and science, the great dependence, — and may I not add, the safe dependence? — as it ever has been, must continue to be upon the bounty of enlightened, liberal, and public-spirited individuals."[40] And so it remained, for a good many years.

What happened in Cincinnati was no isolated incident. Across America, in frontier towns and sophisticated coastal cities, citizens and business leaders — quite without the aid of government — banded together in literary and scientific societies to promote the diffusion not only of useful but even theoretical knowledge. Thomas Bender of New York University, writing in the *History of Education Quarterly*, argues that "the quality of scientific instruction was much higher than many generalizations in American educational history would lead one to expect." After all, as he points out, Alexis de

Tocqueville remarked that a "multitude" of Americans practiced science, and the spread of private "lyceums" in the frontier states testifies to the hunger ordinary Americans had for knowledge.[41] Certainly the percentage of such folks who would not only sit through a 2 or 3 hour lecture on some arcane scientific subject but also *pay for the privilege* would fill any contemporary college professor lecturing to a classroom of bored teenagers with envy.

The problem is that historians have swallowed hook, line, and sinker the scornful picture of an early American scientific scene dominated by vast public indifference, hucksters and charlatans, and a handful of far-seeing men who understood the pressing need for government subsidy if science were to flourish in the new republic. This picture was often painted by those very men who had a highly personal and remunerative interest in attracting such subsidy. As Bender writes, "By accepting their assertions about the indifference of Americans to science, historians have ignored the vigor and pervasiveness of a different kind of science" — the community science of which the Cincinnati Observatory is such a shining example.[42]

The science of the era was decentralized. Each town had its own society or societies dedicated to the spread of knowledge, and they often produced journals of impressive quality.

Cincinnati's eminent homegrown scientists included Daniel Drake, the "physician to the West," who helped to found in that Ohio River city a medical school, a college, a museum, and a scientific journal. Drake was "an urban cultural booster," says Bender, a "community-oriented" man whose type could be found all across the American inland, from Buffalo to St. Louis.[43] The Daniel Drakes tended not to appear in books and papers written by scientists seeking public subsidy: they confused the narrative too much.

Historians of science Daniel J. Kevles, Jeffrey L. Sturchio, and P. Thomas Carroll, writing in *Science*, also dispel the myth that post-Civil War 19th century American science was an "intellectual backwater." In fact, they write, the sciences in America, particularly the life and earth sciences, were the scene of "considerable vitality." For our skewed picture they blame, in part, scientists themselves, especially those who engage in "pure science," asserting that these fearless seekers of The Truth have disparaged their forebears and in so doing "have often used history to argue for the maintenance or, better, the enlargement of R&D support."[44] National pride is a funny thing. Intellectuals often dismiss it as irrational superstition with no relevance to the work of scientists — until scientists want to convince the central state to endow their work, in which case national pride becomes an essential component of a healthy society.

Squeezing money out of Washington was not the goal of the American Association for the Advancement of Science, founded in 1848, and with a membership of professionals and amateurs that fluctuated between 500 and 1,000 members at midcentury.

The association was birthed in Philadelphia out of the Association of American Geologists and Naturalists and came into being with two wings: a natural science section and a physical science section. The first presidents of these sections were, respectively, Louis Agassiz and Joseph Henry, so this was not exactly a scientific desert populated by fugitive yokels. Its annual and biannual meetings reflected the decentralized nature of the American scientific enterprise: the nation's scientists gathered in Charleston, New Haven, Cincinnati, Buffalo, Burlington, and Dubuque. The notion that science was the preserve of a small and esoteric priesthood that could only meet in Manhattan or Boston would have been utterly foreign to the majority of American scientists of the 19th century. In his history of the American Association for the Advancement of Science, Herman Fairchild took pride in noting that "Our association has followed the 'democratic' American custom of throwing all sessions open to the public without charge." By contrast, Fairchild wrote, "The British association has been somewhat exclusive, and from its beginning their scientific and social functions have not been freely open to the public."[45]

The association "worried more about warding off political meddling, religious hostility, and public ridicule than it did about tapping the public's purse or otherwise actively exploiting public favor," notes Robert V. Bruce. "The Association's main concern was not what it could get the public to do for scientists, but what it could help scientists do for themselves."[46] Thus it did not lobby for government grants; instead, it held well-attended meetings, sponsored speakers, published papers, and otherwise encouraged the growth of an American scientific community.

"Men of science... had to act as their own entrepreneurs, availing themselves of every opportunity to secure patronage and general recognition for basic scientific research," wrote Howard S. Miller in *Dollars for Research: Science and Its Patrons in Nineteenth-Century America* (1970). And yet "qualitatively, in the nineteenth century financing from the private sector did much to give tone and tempo to American scientific work." For there remained "a gnawing fear that government support of science was somehow unconstitutional," a fear that was based in fact but faded in the 20th century, the century of the state.

Not every man of science shared that fear. Alexander Dallas Bache, great-grandson of Benjamin Franklin and Superintendent of the U.S. Coast

Survey, told an 1844 scientific congress in Washington that "we want a bounty for research."[47] He would get it — eventually.

This was, in part, the familiar story of a developing priesthood. Young scientists in search of greater status scorned their predecessors as fakes and mountebanks, or as Joseph Henry, first Secretary of the Smithsonian Institution, said, "every man who can burn phosphorous in oxygen and exhibit a few experiments to a class of young ladies is called a man of science."[48] The contempt for those outside the charmed circle is palpable. In the view of Henry and the government scientists, the amateur, the scientist not living in Boston or New York or Washington, DC, was a fraud, a joke, a charlatan.

By 1861, less than 10 percent of the leading American scientists (those who made the *Dictionary of American Biography*) were amateurs, while a majority (60 percent) earned their full income from science, often in colleges.[49] Science as a vocation as well as avocation was coming into its own. Some fields – geology, for instance, thanks to mining — held out the promise (even if it was only irregularly redeemed) of a good living, perhaps even riches. So did chemistry, as American industries from textiles to sugar refining sought better ways of production.

Geological surveys are an early example of government–science collaboration, including at the state level. "The geologists wrote letters in support of one another to governors and legislative committees. They ran survey-puffing editorials in scientific journals," says Robert V. Bruce. They lobbied hard. And successfully, to a point.

Now and then, they stopped to wonder about the costs. As Bruce writes, "Since a cardinal goal of scientists was unfettered research, they had occasional qualms about dependence on government." Yet the scientists gathered around the Smithsonian, for instance, "were shying from the bit, not the feedbag."

As Bruce writes of the scientists of the Coast Survey and the early Smithsonian, "what they wanted was self-rule for science, support without strings."[50] Don't we all? They understood that direct federal subsidy would come with strings that could be tightened like a noose, so they turned much of their attention toward the nation's young universities, which offered employment to scientists, although the teaching loads — stop me if you've heard this one before, and since — were too heavy to get much serious research done.

By 1880, more than 3,000 Americans were employed in scientific enterprises, with chemistry the most populated field and physics the least. Five hundred or so published their research in respected journals, and although these included such vehicles as the *American Journal of Science*, they were just as likely, if not more so, to publish in the volumes of state or local associations — for instance, the *Proceedings of the Connecticut Academy of Sciences*. This "localism," as Kevles, Sturchio, and Carroll term it, was an efficient and decentralized way of conducting research and conveying its results.[51] After all, American governance was, in theory, decentralized, and the country itself was so vast and diverse — as geologists were discovering from the Redwood forests to the Gulfstream waters — that somehow a decentralized science seemed to fit the United States.

"After the Civil War," write the historians of science John Lankford and Ricky L. Slavings, the "dramatic growth of the economy created individuals of great wealth, some of whom became generous patrons of astronomy."[52] If the 1880s were an era known, depending on one's ideological predilections, as the Age of the Robber Barons, the Great Barbecue, or the Age of Entrepreneurship, they also stand as a stellar example of the ways that private assistance can endow not only the lonely scientist at his microscope but also the largest and most spectacular scientific projects of an age.

Louis Agassiz said of the responsibilities of donors: "I hold it one of the duties of those who have the means, to help those who have only their head, and who go to work with an empty pocket."[53] Enough wealthy Americans agreed that the nation's science flourished.

The 19th century, and especially its last quarter, was an age of incomparable richness in the private support of American science, particularly astronomy. Numerous observatories were built without a cent of public monies. Wealthy men, from James Lick to Leander J. McCormick, endowed telescopes and buildings in which to house them.

Writes Joseph P. Martino in *Science Funding: Politics and Porkbarrel* (1992), "astronomy was the big science of the day. Quite clearly, in the late nineteenth and early twentieth centuries, big science not only could be but was funded privately rather than by the government. With the sole exception of the Naval Observatory, every observatory constructed in the United States was built and operated with private funds, and most of them with funds donated by private philanthropists. The end result was that by the third decade of the twentieth century American astronomy led the rest of the world."[54]

"Of all the sciences, astronomy was the most popular object of private support in the United States," says Howard S. Miller.[55] And so there arose talented men who figured out how to match philanthropists, male and female, with deserving projects, institutions, and scientists.

Edward C. Pickering directed the Harvard College Observatory from 1877 until 1919, leading it through an encyclopedic survey of the brightness of all known stars. He was a whiz at attracting private donations, enshrining Harvard as a locus of American astronomy; the observatory's endowment grew from $175,000 to over $900,000 during his tenure.

In an influential address in the first year of his directorship, entitled "Endowment of Research," Pickering disparaged state financing of scientific research, for government subvention would be "lost willfully, or spent ignorantly by the many hands through which the money passes, before it is brought to bear on the scientific conditions of the problem."[56] He proposed a private research institute, which would parcel out needed funds from a pool donated by philanthropists.

In 1886, Pickering set out to raise a $100,000 capital fund, the interest from which he would expend each year in the form of grants to deserving astronomers. It took him four years to find a donor, Catherine Wolfe Bruce, who had earlier been the donor of a 24-inch telescope to Harvard. In 1890, Miss Bruce donated $6,000 for Pickering to disburse. He did so in a scrupulous manner, after sending out circulars to journals of science and astronomers seeking applicants for grants not to exceed $500 apiece. He chose fifteen applicants from a pool of 84.

It was a modest start, but Pickering was not alone in exploring ways in which philanthropic contributions might be matched to worthy scientists. As Howard Plotkin notes in the journal *Isis*, Elizabeth Thompson had established a fund bearing her name in 1885 which subsidized researchers in the broad range of the sciences. In 1883, hardware heiress Mary Anna Draper, who endowed astronomical work at Harvard College Observatory, established The Henry Draper Fund to assist "original investigations in astronomical physics."[57]

The creed underlying these gifts was expressed by Andrew Carnegie, the Scottish-born steel magnate who became the greatest philanthropist in American history. "If any millionaire is interested in the ennobling study of astronomy... here is an example which could well be followed, for the progress made in astronomical instruments and appliances is so great and continuous that every few years a new telescope might be judiciously given to one of the observatories upon this continent, the last being always the largest and best, and certain to carry further and further the knowledge of the universe and of our relations to it here upon the earth."[58]

Carnegie's immediate inspiration was the Lick Observatory on Mount Hamilton, east of San Jose, California. Its patron was perhaps less civic-minded than most great donors, not to mention Mr. Carnegie, but that hardly detracts from his scientific legacy.

James Lick, a wealthy San Franciscan whose fortune was based in real estate, was benefactor of the Lick Observatory. Lick was, to put it mildly, eccentric. Pondering his own mortality, he considered several possible monuments to himself. His original idea for a monument to his own greatness was a huge marble pyramid in his honor, which was to be constructed in downtown San Francisco — though not by slaves. Thanks to George Davidson, president of the California Academy of Sciences, Lick was talked out of the pyramid idea and "was eventually persuaded to endow the biggest telescope in the world so that it could be determined whether there are animals living on the moon."[59] There probably aren't, but it was not as stupid a question as it appears to us nowadays. If curiosity killed the cat (who did not live on the moon), it also led to the building of an observatory that contained a 36-inch refractor telescope — the world's largest at the time — whose lens was ground by the great Alvan Clark of Massachusetts, the premier optician of the age. Lick preferred that the telescope, like the would-be pyramid, be located in downtown San Francisco, but he was eventually convinced that the mountains, and not the city streets, was the place for his telescopic monument.

James Lick was perhaps not a Galileo-like seeker of truth, but whatever his motives, he pledged $700,000 to the fund for construction, which was overseen by 31-year old Richard Floyd, who had no experience whatsoever in the world of the stars, though he had been a sea captain. Floyd navigated his way past the shoals of fierce internal politics, and in 1888 the Lick Observatory opened its doors and its roof to first light.[60] Its star astronomer, Edward Emerson Barnard, discovered a moon of Jupiter and was widely regarded as among the world's finest observational astronomers.

The man at the center of much of the private telescope and observatory building of the late 19th and early 20th centuries was George Ellery Hale: son of a Chicago elevator manufacturer, MIT graduate, founder of the *Astrophysical Journal*, cofounder of the American Astronomical and Astrophysical Society, electee at age 22 to the Royal Astronomical Society, and persuader of men of vast fortunes to direct philanthropy toward the science of astronomy. From his perch at the University of Chicago, Hale was instrumental in persuading Charles Yerkes to immortalize himself through an observatory on the shore of Lake Geneva, Wisconsin.

Charles Yerkes was a freebooter who never took his eye off the main chance, an ex-convict who tried to establish a monopoly on streetcars in Chicago in the late 19th century. He was as honest as the day is short, and though many of the buccaneers and financiers of the day had a certain swagger and style about them — as well as concrete accomplishments to their credit — Yerkes was probably the most hated man in Chicago. George Ellery Hale, who knew a thing or two about human nature as well as stellar nature, convinced Yerkes that endowing a big telescope would do wonders for his image. He would, overnight, transform himself from Yerkes the Crook to Yerkes the Philanthropic Man of Science. Or so Hale told him. For his part, Yerkes told Hale, "You shall have all you need if you'll only lick the Lick!"[61]

Hale had convinced the street car baron of Chicago to gift the University of Chicago with a refractor which would make the Yerkes Observatory "a Mecca of thousands of science-loving pilgrims, as the Lick Observatory, even in its isolated position, is to-day. And the donor could have no more enduring monument." Even the otherwise critical Chicago *Times-Herald* admitted, "Whatever opinion we may hold of Mr. Yerkes in his relations with the people of Chicago, there can be only one opinion, and that extremely complimentary, of Mr. Yerkes as the founder of the Observatory at Lake Geneva."[62]

In 1897 Hale licked the Lick, if barely: the showpiece scope of the new Yerkes Observatory, for which the namesake put up $300,000, was the 40-inch refractor, just four inches larger than the Lick's largest telescope, a superb instrument also made by Alvan Clark. (The firm of Alvan Clark & Sons would design many of the finest telescopes of the age, paid for by private individuals wishing to contribute to the fund of human knowledge. For instance, heiress Catherine Wolfe Bruce donated $50,000 to construct a 24-inch Clark refractor for the Harvard College Observatory; it saw first light in 1893 and two years later was moved to Harvard's observing station at Arequipa, Peru. Bruce also sponsored dozens of researchers to the tune of almost $175,000 over the last decade of the century of philanthropy. She even paid the salary of an astrophysicist at Yerkes when the observatory's eponym refused to pony up.[63])

Having established the Yerkes Observatory as the jewel of the astronomical Midwest, Hale looked for clearer skies and even fatter bank accounts. Explains Daniel J. Kevles, "The businessman's son got along well with men of wealth and was naturally at ease marshalling arguments for science to a wide variety of powerful people."[64] He used these talents plentifully in his next project: the Mount Wilson Observatory outside Los Angeles.

Hale understood the implications of Andrew Carnegie's philanthropy. He approached the steel titan's Carnegie Institution, which was founded in 1902 with a gift of $10 million from its namesake and a deed of trust whose first adjuration was "to promote original research, paying great attention thereto as one of the most important of all departments."[65] Hale asked Carnegie for money — lots of it. Carnegie responded. As he responded to other astronomers, it should be noted. Astronomical research accounted for 40 percent of the requests for scientific aid made to the Carnegie Institution between January 1902 and October 1903, its first year and three-quarters of operation; Carnegie "was central to the expansion of American astronomy," write Lankford and Slavings. In fact, the elevated place of astronomy among the sciences was something of a sore point for other disciplines: about half of the money disbursed to science through philanthropic endowments in 1900 was for the study of the stars.[66]

Carnegie wanted his institution to "seek the exceptional man, and aid him."[67] Hale was exceptional, and Carnegie aided him.

Mount Wilson Observatory was born in 1904. Its first major telescope was a 60-inch reflector that went into service in 1908 and was subsidized by Carnegie, whose generosity underwrote the observatory. Hale, who seems to have had a serious case of size envy, immediately set to work trying to entice a donor to pay for a much larger instrument: a 100-inch reflector, largest in the world by far. John D. Hooker, a Los Angeles iron works executive, was the man who made possible Mt. Wilson's star instrument, the 100-inch reflector, with his donation of $100,000. Its first light was admitted in 1917, as the world was going dark in the Great European War.

Hale was not satisfied to rest even on this considerable laurel. If a 100-inch telescope was viable, how about a 200-inch reflector? Skeptics abounded. "The Bureau of Standards said such a telescope was probably impossible," say film documentarians Todd and Robin Mason, makers of *The Journey to Palomar*. It seemed like an act of hubris, at best, yet Hale had a track record, and so another capitalist-endowed organization, the Rockefeller Foundation, put up $6 million for the project in 1928.

The 200-inch telescope — the "Perfect Machine," as it was dubbed — at Palomar Observatory on Palomar Mountain near San Diego, California, would take two decades to complete. The Corning Glass Works in New York made the mirror of Pyrex in what was a long, slow, arduous process marked by trial, error, and finally triumph. In 1936, the 200-inch mirror traveled by train across the country "at a stately speed of 25 miles per hour" in what the California historian Kevin Starr called the "moon-shot" of the mid-1930s.[68] But the critical difference is that while the moon-shot was a triumph of government science, the Palomar telescope was the last major achievement of the glorious era of privately funded astronomy. The United

States had become the center of astronomical science; it hosted the biggest telescopes, the most productive observatories, the top astronomers. And it had been done, almost entirely, with private-sector money. John Quincy Adams had gotten it wrong.

The Hale Telescope, as the 200-inch instrument is called, saw first light in January 1949. It was the largest telescope in the world for almost 50 years, before finally being displaced from that perch by the 400-inch Keck I Telescope in Hawaii, which was funded by the W.M. Keck Foundation, a charitable foundation fueled by money from the Superior Oil fortune. The Keck Telescope was the result of "the largest private gift ever," according to *Science* magazine[69] — a gift from the W.M. Keck Foundation to Cal Tech. (In December 1975, the BTA-6 238-inch reflector saw first light in the Soviet Union, but its mirror was plagued by so many imperfections that it was essentially inoperative and therefore can't be counted.) The Masons, the documentarians of Palomar, asked an engineer how long the Hale Telescope could be in service. "Forever," he replied.[70] Not a bad legacy for George Ellery Hale.

A Department of Science?

Despite the extraordinary strides made by privately funded American science, some politicized scientists and scientifically minded politicians clamored for a seat at the Cabinet table. The din grew loudest just as James Lick, Charles Yerkes, and George Ellery Hale were showing the world new vistas in privately financed science.

A joint commission (the Allison Commission) appointed by Congress in 1884 had actually recommended the consolidation of the scattered science-related bureaus of the government and the establishment of a Department of Science, an idea that went nowhere but that did shed some light on the changing landscape.

In a way, the debates of the 1880s represent the last time that philosophical objections to the government subsidy of science carried enough political heft to win the day.

Not that subsidy didn't have its partisans.

Some encouraged the United States to join the European nations in patronizing the work of scientists. *Science* magazine, in October 1885, quoted George Washington: "Promote as an object of primary importance institutions for the general diffusion of knowledge. In proportion as the structure of a government gives force to public opinion, it is essential that public opinion should be enlightened." Washington's words can be taken as an endorsement of federal support of museums or universities, but *Science*, edited, evidently, by very loose constructionists, took them to mean that science was a primary concern of statecraft. The magazine pointed to such federal offices as the U.S. Fish Commission as evidence that in "some respects... this young country is in advance of all European states in joining science to its administrative offices."[1]

That the nascent bureaucracy of the United States should be compared to that of the European nations — favorably! — might seem inconsistent with the general American distaste for statism. And in fact in a contemporaneous issue of *The Nation*, which today we think of as a left-wing publication but which in the 19th century was the leading American journal of classical-liberal opinion, marine biologist Alexander Agassiz of the Harvard Museum of Comparative Zoology, son of the famed paleontologist Louis Agassiz, laid out the case against a burgeoning science bureaucracy.

Affairs of science had never been regarded as the province of the U.S. government, noted Agassiz. Exceptions were rare and well-defined: the U.S. Naval Observatory, whose task was not so much astronomical research as it was aiding in navigation; the Nautical Almanac Office, whose purposes were similar; and the Smithsonian Institution, borne of a private bequest. Territorial expansion and the ambitions for a transcontinental railroad spurred Washington into sponsoring work in geology and biology as well as agriculture, coastal and mining surveys, and the like, so that by the 1880s "the scientific activity at Washington has been something prodigious."

This activity, which of necessity was overseen by government bureaucrats, threatened the independence of men of science, argued Agassiz. "We cannot expect that men of eminent scientific attainments will submit to such regulations," he told the readers of *The Nation*, "or will serve effectively under chiefs selected merely for their administrative capacity." He worried that the administrative drones at the head of these departments would be out of sympathy with the needs of science, and that placing the departments under military men would lead to an even worse situation, in which scientists became regimented servants of the state.

Yet to place scientists in charge of these bureaus carried its own set of dangers, for "[u]nfortunately, men of science have frequently shown little aptitude for administrative work." Catching the scent of federal dollars, they have forgotten "the dictates of common sense." For "they have incurred expenses with which the results obtained are by no means commensurate.

They have encouraged a class of men clever in scientific cant, and ready to encourage any scheme as long as it was paid from the Government purse."

Federal subsidy, in other words, turned clever men of science into wily beggars, or shrewd manipulators of political levers and public emotions who knew how to pry open the public till — a practice they would learn with surpassing skill in the next century.

Agassiz warned that the "friends of a paternal government would like to see the science of the country centralized, and the work of the bureaus gradually absorbing all the best available men in their respective departments, making Washington a great scientific centre." The cost of this would be an atrophy of American science, as centralized administration always "interfere[s] with the development of individuality."

Thankfully, the early push for a national university had failed, for as Agassiz wrote, while at "first glance, the idea of a great university established at Washington in connection with the scientific bureaus, and backed by Government resources, is dazzling," upon closer inspection the dazzle fades and hard questions intrude. Such as: "is it wise for the Government to enter directly as a competitor in the field of higher education, except in the training of army and naval officers?" Is it even conceivable that "the administration of such an institution could be kept out of politics"?

Agassiz is doubtful. For the "past history of our scientific bureaus has been such as to suggest nothing but disaster from the centralization of science at the capital." The "present capacity of some of our bureaus for indefinite expansion by constantly encroaching upon the field of individual activity" bodes ill for the independence of scientists on the government dole. Agassiz will concede to Washington certain limited powers, but the "Government should limit its support of science to such work as is within neither the province nor the capacity of the individual or of the universities, or of associations and scientific societies."[72] It should set up no schools, no centralized laboratories, no bureaus beyond the pale of constitutional governance. Thus, in 1885, speaketh The Nation. Some things do not get better with age.

Rep. Hilary Herbert (D-AL) insisted in 1886 that "Government patronage shackles that spirit of independent thought which is the life of science."[73] Looking around at a national scientific establishment that, though skeletal by today's standards, encompassed the Naval Observatory, the Coast and Geodetic Survey, and an incipient weather service, Herbert charged that "the United States government was extravagantly investing more annually in scientific research than all the nations of western Europe combined."[74]

Rep. Herbert explained himself thusly: "I am radically Democratic in my views. I believe in as little government as possible — that Government should keep its hands off and allow the individual fair play. This is the doctrine I learned from Adam Smith & Buckle, from Jefferson, Benton and Calhoun, and from this standpoint I believe we have too much to do (the Gov't) with pure science."[75] He gave the Western explorer John Wesley Powell a hard time as a leading member of the Allison Committee, and while Powell's U.S. Geological and Geographical Survey survived, the proposed Cabinet-level Department of Science did not.

"Government science before the Civil War was largely, though not quite exclusively, Joseph Henry and Spencer Baird of the Smithsonian," wrote the great Western novelist and biographer Wallace Stegner. "Geology was a States' rights matter, topography and mapping were diversions to occupy the peacetime Army, time and weather were for the Navy to play with, and too much of private science was the occupation of amateurs."[76] Yet Stegner, an environmentalist, would later have cause to wonder if in fact the flurry of postwar governmental bodies — the Geological Survey, the National Park Service, the Forest Service, the Bureau of Mines, the Reclamation Service, and others — was quite the unadulterated good it may have appeared to forward-thinking progressive men of the era, such as his subject, the brave explorer of the West, John Wesley Powell of the United States Geological and Geographical Survey of the Rocky Mountain Region.

Given the power, the federal government would spend taxpayers' dollars on dams that despoiled the natural beauty of the region, and in the sharpest insult, it flooded Glen Canyon and named the resultant manmade body of water Lake Powell. It would subsidize mining companies and ranchers; and it would consign stretches of the American desert to the U.S. military for weapons testing. If Powell and Stegner cared about the West, as they most certainly did, they mistook an enemy for a friend.

Besides, were private science, state geological surveys, and a Navy that occupied itself with chronometry rather than sailing to the Philippines really all that baneful? Alabama Democratic Representative Hilary Herbert, Powell's nemesis, pestered him in 1884–1885 congressional investigations for wasting too much money on topographical surveys. Powell was map-happy, charged Herbert, and if we laugh at Herbert as a doltish pennyp-incher, as the history books do, we might listen to Alexander Agassiz, who concurred with Herbert in believing, in Stegner's words, that "economic geology should be left to the mining companies, paleontology to the universities and private individuals. He saw no reason why scientists should ask more of the government than literary men or artists or any other of the learned professions."[77]

Agassiz cannot be written off as a Southern bumpkin, or an obscurantist know-nothing. He was an eminent man of science, bearer of one of the great surnames in American intellectual history, and he denied that science was the province of government.

Wallace Stegner saw the subsidization of science as having much broader implications than merely providing gainful employment for geologists. "The concept of the welfare state edged into the American consciousness and into American institutions more through the scientific bureaus of government than by any other way.... It began as public information and extended gradually into a degree of control and paternalism increased by every national crisis and every step of the increasing concentration of power in Washington. The welfare state was present in embryo in Joseph Henry's Weather Bureau in the eighteen-fifties."[78]

The development of government meteorological bureaus, as with other scientific activities of the state, was boosted greatly by the Civil War. As Eric R. Miller wrote in a history of U.S. meteorological institutions, the study of weather "was scantily supported by eking out small sums from budgets intended for other purposes, before the war, and then afterward [it was marked by] the quick development of a single relatively lavishly supported institution."

In the early decades of the republic, records of the weather were kept by the Army Medical Department and the Interior Department, though these were desultory and hardly comprehensive. Such work was regarded as the province of the states, and in fact, New York and Pennsylvania, among others, supported the purchase of barometers and thermometers and rain gages and their employment by designated agents. The Smithsonian coordinated hundreds of weather observers before the Civil War, and after the war, the Signal Service undertook weather observations in some earnest. Still, it wasn't until 1895 that the various state weather services were, in effect, incorporated into (or disappeared as a result of the creation of) the United States Weather Bureau, a civilian agency.[79]

Despite the "handicap" of a lack of government financing, American astronomy made extraordinary strides in the 19th century. Indeed, as Stephen Brush of the University of Maryland explained in "The Rise of Astronomy in America," the late 19th century, which has suffered almost "complete neglect" by historians of American science (who tend to concentrate on later eras), was a period in which the U.S. rose to "world leadership in a relatively 'pure' branch of science... well before the immigration of

European scientists in the 1930s, which is usually credited with producing our postwar superiority in fundamental research."[80]

Referencing books by non-Americans to avoid any hint of nationalist bias, Brush compiled a list of the ten major astronomers and major events in astronomy for three half-century periods: 1801–1850, 1851–1900, and 1901–1950. As he had expected, Germans dominated the science in the first half of the 19th century. In fact, not a single American appears as either a major astronomer or a participant in a major astronomical event in that period. Even so, the free and unregimented world of American science was starting to produce significant figures of widely varied backgrounds. There was Benjamin Banneker, the African-American almanac publisher of the early republic; Alvan Clark, telescope-maker extraordinaire of Massachusetts, who with his son would make some of the greatest optical instruments of that or any other age, including the lead telescopes for the Lick and Yerkes Observatories; and Maria Mitchell, the Yankee skygazer who would teach astronomy at Vassar. These and others "were self-educated in astronomy," writes Brush. They "worked alone and had no institutional support until they established their reputations."

By the second half of the century, American astronomy, almost completely unsubsidized by the government, had achieved a status equaled only by that of Britain. Americans Edward C. Pickering, Simon Newcomb, and Edward Emerson Barnard grace the top ten list; Americans figure in six of the ten major events in astronomy. In fields requiring doggedness and sheer hard work — for instance, the discovery of asteroids — Americans excelled. American astronomy of the 19th century was marked by "democratic *openness* to bright, energetic people without professional training or certification by the establishment and an emphasis on *practical skills and technology*."[81]

They were bluebloods and sons of the frontier alike. Consider Percival Lowell.

The Boston aristocrat Lowell, though undeniably an eccentric and a man who was, apparently, dead wrong on his magnificent obsession — the question of life on Mars — shows us an alternative track down which scientific showmen might go. Rather than scaring the bejeebers out of the public in the hope of a windfall from the national treasury, Percival Lowell relied not just on inherited wealth but on his own labor — best-selling books explaining his theories of Martian life — to support scientific research.

In 1894 he built, with his own fortune, Lowell Observatory in Flagstaff, Arizona, in the state that has perhaps the best "seeing" in America. He chose his spot wisely. Although the astronomers he hired on his own dime (or millions of dimes) were not able to detect evidence of life on Mars, the Lowell Observatory's research was not all for naught. Long convinced that a "Planet X" must exist beyond Neptune, Percival Lowell made its detection a prime mission of the Lowell Observatory. In 1930, 14 years after Lowell's death, a Lowell Observatory employee named Clyde Tombaugh found "Planet X" — or at least *a* "Planet X," if not *the* "Planet X" — and named it Pluto, the now-disputed ninth planet of the solar system. Fittingly, the only planet discovered by an American was tracked down with purely private money.

Gilded Age America also became the home of the great comet-hunters. Chief among them was Lewis Swift (1820–1913), a farmer and hardware-store owner from Monroe County in Upstate New York who discovered thirteen of the sidereal messengers between 1862 and 1899.

(One of his discoveries, Comet Swift-Tuttle, later contributed to the Armageddon from the Skies craze of the late 20th century when in 1992 the estimate that "Swift-Tuttle had a 1-in-10,000 chance of colliding with the Earth" in 2126 merited a *New York Times* headline blaring "Scientists Ponder Saving Planet from Earth-Bound Comet."[82] Of course it turned out that Swift-Tuttle was not Earth-bound, though it served its purpose in fanning the panic from which federal subsidies flow.)

Swift was that classic American blend of smarts and get up and go. He taught himself astronomy and built a series of makeshift rooftop observatories. One such, Professor Peter T. Wlasuk of Florida International University relates in the *Quarterly Journal of the Royal Astronomical Society*, was "on a platform atop the roof of Duffy's Cider Mill, on White Street near the Genesee River in Rochester. Icy weather often compelled Swift to crawl 100 ft across the cider mill's slanted roof on hands and knees."

His story proved irresistible to the press, which lauded Lewis Swift as the "People's Astronomer." He lectured (gratis), he threw huge star parties to which the hoi polloi were invited, and his enthusiasm for the science of astronomy spread far beyond the bounds of his town. Bausch & Lomb took him on as an optical consultant, though he was, admittedly, an imperfect judge in matters of the eye. As Wlasuk relates, Swift was convinced that the curveball was an optical illusion. He strode to the plate to test his hypothesis. "The startled astronomer quickly recanted his opinion," says Wlasuk, "after almost being hit on the head by one of the pitcher's 'illusions'!"

Swift's exploits attracted the attention of a wealthy Rochesterian, Hulbert Harrington Warner, whose millions had been earned, in part, through an obvious item of medical quackery called "Safe Liver Pills." Warner paid for the construction of a "magnificent" observatory that bore his name and in which Swift would scan the skies for comets. By 1882, Swift was "a full-time professional astronomer" using a 16-inch Alvan Clark refractor, which was "the fourth largest telescope in the United States at that time." Swift remained the People's Astronomer, opening the Warner Observatory to the public on designated nights.

"The Warner Observatory is distinctively a private institution," declared Lewis Swift, and for ten grand years Messrs. Swift and Warner combined to make this private, populist entity a nationally recognized center for the exploration of comets and nebulae. When Mr. Warner lost his fortune in the depression of 1893, the facility was shuttered, and Swift moved west, to Echo Mountain near Los Angeles, California, where he and the philanthropist Thaddeus Lowe, who had invented a better way of making artificial ice, launched the Echo Mountain Observatory.

"Lewis Swift was one of the last of his kind," writes Peter T. Wlasuk — "an amateur astronomer who, by availing himself of wealthy patrons, was able to convert his avocation into a vocation."[83]

Yet even as instruments got bigger, and telescopes atop cider mills became less central to astronomical discovery, privately funded American science prospered in comparison with the Europeans, who were more reliant upon subsidy from state or king.

The first half of the 20th century on Stephen Brush's list is dominated by Americans: George Ellery Hale, Harlow Shapley, Henry Norris Russell, Edwin Hubble, the German-American Walter Baade, and the Dutch-American Gerard Kuiper. Americans are key participants in eight of the ten major astronomical events, and the man responsible for the ninth, Albert Einstein, would immigrate to the United States. Such milestones as the construction of large telescopes and the birth of radio astronomy were almost purely American phenomena.

American dominance in astronomy cannot be attributed in any way to the Second World War, which cast such a shadow over the scientific programs of many nations. "The discoveries announced by 1930 were sufficient to put the United States ahead of all other countries," writes Brush, and this was the era before state patronage of astronomy. As Brush writes, "By 1930, American astronomy had risen as far as it could: to the top."[84]

Note that this was prior to the New Deal, prior to the Second World War and the Cold War, prior to the vast reconfiguration of American government that occurred in the middle part of the 20th century. American astronomy had come from almost literally nowhere to be the best in the world in a century, and it had done so because of private philanthropy, private institutions, and privately employed astronomers. This triumph was in no way, shape, or form a tribute to state action or subsidy: it was a testament to the labors and genius of an American science nursed and cultivated in the private sector.

Even as fervent an advocate of federally subsidized science as Jerome Wiesner, Special Assistant for Science and Technology to Presidents Kennedy and Johnson, conceded that the "development of electric power, electric light, railroads, telephone, radio, television, medicine, automobile, automation in industry and other labor-saving devices were all creations of private initiative."[85] Wiesner and the Kennedy administration actually opposed the creation of a Department of Science, which since the 1880s had floated in and out of legislative hoppers. They feared that centralizing science in the bureaucracy would also isolate it.

Joseph P. Martino observes that "with the coming of federal funding after World War II, private philanthropy was almost completely driven out of science." He continues, "It seems clear that if the federal government had not taken over the support of science, the amounts forthcoming from philanthropy would be even greater than they are now."[86]

By the midpoint of the 20th century, astronomers had become better acquainted with the public sector, as they were using "new instruments (often developed for military purposes) and [were] funded by new federal patrons (including both the military and the National Science Foundation)."[87] Private observatories and their telescopes endured — they were built to last, and many still did essential work — but Washington was becoming, to many, the center of the universe.

World War I had ramped up the martial applications of scientific research, and not only through such bureaus as the Chemical Warfare Service. As George Ellery Hale remarked hopefully, "war should mean research."[88]

"World War I had profound effects on every part of American science, whether supported by the government, by the universities, or by the foundations," writes A. Hunter Dupree.[89] It even gave the NAS something to do. As Obama chief of staff Rahm Emanuel would later quip, never waste a crisis. That is a lesson that advocates of Big Science have learned well.

In 1916, the National Academy of Sciences created a council to serve as a kind of ad hoc advisor on military applications of science. Two years later, in 1918, President Woodrow Wilson created, by executive order, the National Research Council as an umbrella under which representatives of industry, the university, and government cooperated in "the mobilisation [sic] of American science for national defense."[90] This coalition in some ways marked a sea-change in the relation of American science and government; as Kevles writes, "University and industrial laboratories, historically remote from the government's needs, were working on military problems" in an "unprecedented and fruitful collaboration of university and industrial scientists with the military."[91] From chemical warfare to submarine detection technology, science was serving the war effort. Indeed, the "wartime NRC [National Research Council] became a central scientific agency to an extent never dreamed of by the National Academy."[92]

George Ellery Hale, impressed by the wartime cooperation of science and state, "set out to make the coalition permanent." Puffed up with that sense of self-importance that high-level government work can give to men — sure, scientific discovery has its thrills, and 100-inch telescopes are a kick, but what compares to ordering men around and issuing portentous pronouncements on papers with official state watermarks? — Hale drafted an executive order for President Wilson to sign which would have made the NRC a more or less permanent agency of government with the responsibility for subsidizing a wide variety of scientific research, endowing research fellowships, overseeing the expanding scientific demands of the federal government, and otherwise turning on its head the traditional relationship of state and science. This crisis was not going to be wasted. As Daniel Kevles notes, Hale disparaged the Democratic Congress as a band of know-nothings who cared more about "the needs of 'the Little Red School House on the hill'" than the demands of important scientists. Therefore, an executive order, which made an end run around Congress, was his preferred instrument.

Unfortunately, for Hale, the President sent the proposed executive order out for comment. Secretary of Agriculture David Houston, a political scientist, protested that it raised "questions of constitutional authority."[93] What an old fuddy-duddy Houston was! The order was revised, or watered down, in Hale's estimation, draining power from the National Research Council, and though Wilson did issue it on May 10, 1918, the postwar NRC fizzled, despite gifts from the Carnegie Corporation and the Rockefeller Foundation. The Harding administration, picking up after the spendthrift Wilson White House, preferred economy in government to searching out new tasks for Washington to perform. And curiously, Hale was of two minds on the matter. While he wanted a powerful NRC, he preferred funding to come from private sources — a hybrid that was incoherent in conception and could not thrive without a foreign threat or domestic panic to fan the flames of support.

Fear not: threats and panics were on their way.

Notes

1. James Madison, *Notes of Debates in the Federal Convention of 1787*, with an introduction by Adrienne Koch (Athens: Ohio University Press, 1984/1840), p. 477.

2. *Annals of Congress*, Gales & Seaton's History, May 3, 1790, 1st Congress, 2nd session, p. 1604.

3. *Annals of Congress*, February 3, 1796, 4th Congress, 1st session, p. 288.

4. Quoted in A. Hunter Dupree, *Science in the Federal Government: A History of Policies and Activities to 1940* (Cambridge, MA: Harvard University Press, 1957), p. 9.

5. Robert V. Bruce, *The Launching of Modern American Science 1846–1876* (New York: Knopf, 1987), pp. 36, 76.

6. Charles O. Paullin, "Early Movements for a National Observatory, 1802–1842," *Records of the Columbia Historical Society*, Washington, DC, Vol. 25 (1923), p. 38.

7. Ibid., pp. 39, 41.

8. Ibid., p. 44.

9. *Inaugural Addresses of the Presidents of the United States*, John Quincy Adams, March 4, 1825 (Washington, DC: U.S. Government Printing Office, 1961), p. 47.

10. John Quincy Adams, First Annual Message, December 6, 1825, The American Presidency Project, www.presidency.ucsb.edu.

11. Paullin, "Early Movements for a National Observatory," p. 44.

12. Dupree, p. 42.

13. Paullin, "Early Movements for a National Observatory," pp. 55–56.

14. John Lankford and Rickey L. Slavings, "The Industrialization of American Astronomy, 1880–1940," *Physics Today* (January 1996), p. 35.

15. Nina Burleigh, *The Stranger and the Statesman: James Smithson, John Quincy Adams, and the Making of America's Greatest Museum: The Smithsonian* (New York: HarperCollins, 2003), p. 168.

16. Paullin, "Early Movements for a National Observatory," p. 47.

17. Quoted in Howard S. Miller, *Dollars for Research: Science and Its Patrons in Nineteenth-Century America* (Seattle: University of Washington Press, 1970), p. 11.

18. Paullin, "Early Movements for a National Observatory," pp. 47, 48.

19. *The Smithsonian Institution: Documents Relative to Its Origin and History*, edited by William J. Rhees (Washington, DC: Smithsonian Institution, 1879), pp. 432–33, 438–39, 456.

20. P. Clark, *The Atlantic Monthly*, Vol. 31 (April 1873), p. 471.

21. John Weld Peck, "Symmes' Theory," *Ohio History*, Vol. 18 (1909), p. 34.

22. Ibid., p. 30.

23. Ibid., p. 31.

24. Robert F. Almy, "J.N. Reynolds: A Brief Biography with Particular Reference to Poe and Symmes," *Colophon*, Vol. 2 (Winter 1937), p. 228.

25. Peck, "Symmes' Theory," pp. 38, 28.

26. Almy, "J.N. Reynolds: A Brief Biography with Particular Reference to Poe and Symmes," p. 237.

27. Bruce, *The Launching of Modern American Science 1846–1876*, p. 27.

28. Ibid., p. 297.

29. Dupree, *Science in the Federal Government: A History of Policies and Activities to 1940*, p. 152.
30. Francis G. Caffey, "A Brief Statutory History of the United States Department of Agriculture," *Case and Comment*, Vol. 22 (1916), pp. 723–24, 725.
31. Morrill Act, U.S. Statutes P.L. 37–108.
32. Dupree, *Science in the Federal Government: A History of Policies and Activities to 1940*, p. 151.
33. *A History of the First Half-Century of the National Academy of Sciences 1863–1913*, edited by Frederick W. True (Baltimore: Lord Baltimore Press, 1913), p. 17.
34. Daniel J. Kevles, "George Ellery Hale, the First World War, and the Advancement of Science in America," *Isis*, Vol. 59, No. 4 (Winter 1968), p. 430.
35. Miller, *Dollars for Research: Science and Its Patrons in Nineteenth-Century America*, p. 33.
36. Stephen Goldfarb, "A History of the Cincinnati Observatory, 1842–1872," *Ohio History*, Vol. 78 (1969), p. 173.
37. Ibid., p. 172.
38. Ibid., pp. 173–75.
39. Miller, *Dollars for Research: Science and Its Patrons in Nineteenth-Century America*, pp. 35, 39.
40. Edward Everett, *The Uses of Astronomy* (New York: Ross & Tousey, 1856), pp. 18–19.
41. Thomas Bender, "Science and the Culture of American Communities: The Nineteenth Century," *History of Education Quarterly*, Vol. 16, No. 1 (Spring 1976), pp. 64, 67, 68.
42. Ibid., p. 66.
43. Ibid., p. 65.
44. Daniel J. Kevles, Jeffrey L. Sturchio, and P. Thomas Carroll, "The Sciences in America, Circa 1880," *Science*, Vol. 209, No. 4 (July 1980), p. 27.
45. Herman L. Fairchild, "The History of the American Association for the Advancement of Science," *Science*, Vol. 59, No. 1531 (May 2, 1924), p. 389.
46. Bruce, *The Launching of Modern American Science 1846–1876*, p. 259.
47. Miller, *Dollars for Research: Science and Its Patrons in Nineteenth-Century America*, pp. ix, xi, 5.
48. Ibid., p. 7.
49. Bruce, *The Launching of Modern American Science 1846–1876*, p. 135.
50. Ibid., pp. 167, 168, 169, 225.
51. Kevles, Sturchio, and Carroll, "The Sciences in America, Circa 1880," pp. 27–28.
52. Lankford and Slavings, "The Industrialization of American Astronomy, 1880–1940," p. 34.
53. Quoted in Helen Wright, *James Lick's Monument: The Saga of Captain Richard Floyd and the Building of the Lick Observatory* (Cambridge: Cambridge University Press, 2003/1987), p. 8.
54. Joseph P. Martino, *Science Funding: Politics and Porkbarrel* (New Brunswick, NJ: Transaction, 1992), p. 213.
55. Miller, *Dollars for Research: Science and Its Patrons in Nineteenth-Century America*, p. 112.

56. Howard Plotkin, "Edward C. Pickering and the Endowment of Scientific Research in America, 1877–1918," *Isis*, Vol. 69, No. 1 (March 1978), p. 45.

57. Ibid., pp. 46–47.

58. Stephen Brush, "The Rise of Astronomy in America," *American Studies*, Vol. 20 (1979), p. 54.

59. Ibid., p. 52.

60. See Helen Wright, *James Lick's Monument: The Saga of Captain Richard Floyd and the Building of the Lick Observatory*.

61. Quoted in Brush, "The Rise of Astronomy in America," p. 54.

62. Miller, *Dollars for Research: Science and Its Patrons in Nineteenth-Century America*, pp. 108, 110.

63. Ibid., p. 114.

64. Kevles, "George Ellery Hale, the First World War, and the Advancement of Science in America," p. 428.

65. Plotkin, "Edward C. Pickering and the Endowment of Scientific Research in America, 1877–1918," p. 50.

66. Lankford and Slavings, "The Industrialization of American Astronomy, 1880–1940," p. 39.

67. Plotkin, "Edward C. Pickering and the Endowment of Scientific Research in America, 1877–1918," p. 52.

68. Todd Mason and Robin Mason, "Palomar's Big Eye," *Sky & Telescope* (December 2008), pp. 38–39.

69. Martino, *Science Funding: Politics and Porkbarrel*, p. 304.

70. Mason and Mason, "Palomar's Big Eye," p. 41.

71. "Science and the State," *Science*, Vol. 6, No. 141 (October 16, 1885), p. 326.

72. Alexander Agassiz, "The National Government and Science," *The Nation*, Vol. 41 (December 24, 1885), pp. 525–26.

73. Quoted in Dupree, *Science in the Federal Government: A History of Policies and Activities to 1940*, p. 228.

74. Kevles, Sturchio, and Carroll, "The Sciences in America, Circa 1880," p. 209.

75. Quoted in Miller, *Dollars for Research: Science and Its Patrons in Nineteenth-Century America*, p. 144.

76. Wallace Stegner, *Beyond the Hundredth Meridian: John Wesley Powell and the Opening of the West* (Boston: Houghton Mifflin, 1953), p. 117.

77. Ibid., p. 291.

78. Ibid., p. 334.

79. Eric R. Miller, "The Evolution of Meteorological Institutions in the United States," *Monthly Weather Review*, Vol. 59, No. 1 (January 1931), pp. 1–6.

80. Brush, "The Rise of Astronomy in America," p. 43.

81. Ibid., pp. 50–51.

82. Peter T. Wlasuk, "'So Much for Fame!': The Story of Lewis Swift," *Quarterly Journal of the Royal Astronomical Society*, Vol. 37 (1996), p. 683.

83. Ibid., pp. 688, 691, 694, 695, 696, 705.

84. Brush, "The Rise of Astronomy in America," p. 58.

85. Jerome B. Wiesner, *Where Science and Politics Meet* (New York: McGraw-Hill, 1965), pp. 47–48.

86. Martino, *Science Funding: Politics and Porkbarrel*, pp. 302, 305.

87. Lankford and Slavings, "The Industrialization of American Astronomy, 1880–1940," p. 40.

88. Quoted in Stuart W. Leslie, *The Cold War and American Science: The Military-Industrial-Academic Complex at MIT and Stanford* (New York: Columbia University Press, 1993), p. 4.

89. Dupree, *Science in the Federal Government: A History of Policies and Activities to 1940*, p. 323.

90. Lance E. Davis and Daniel J. Kevles, "The National Research Fund: A Case Study in the Industrial Support of Academic Science," *Minerva*, Vol. 12 (April 1974), p. 208.

91. Kevles, "George Ellery Hale, the First World War, and the Advancement of Science in America," p. 431.

92. Dupree, *Science in the Federal Government: A History of Policies and Activities to 1940*, p. 323.

93. Kevles, "George Ellery Hale, the First World War, and the Advancement of Science in America," pp. 432–33.

Chapter 3

Dr. Bush Fathers a Foundation

World War I, with its poison gases, had been the chemists' war; World War II, with its radar and atom bombs, would become known as the physicists' war. And as Paul Forman of the Smithsonian Institution has quoted physicist Jerrold Zacharias, "World War II was in many ways a watershed for American science and scientists. It changed the nature of what it means to do science and radically altered the relationship between science and government."[1]

If science and scientists, particularly physicists, are sometimes given credit for winning the Second World War for the Allies — at least by those who speak out of earshot of combat veterans — then the federal government returned the favor, many fold. For the war was the turning point in the hitherto uneasy, even detached relationship between science and state. After the war, science would become increasingly dependent on the federal government — and when scientists and science administrators wanted additional funding, they knew what buttons to push, what panics to exploit, what were the magic words that opened the appropriations tap. Even universities, while they would remain nominally independent, became ensnared — quite willingly, for the most part — in the web of federal funding and the federal control that followed.

During the New Deal, the Science Advisory Board of the National Academy of Sciences had recommended that the federal government provide grants-in-aid to academic and industrial scientists pursuing work that may not have practical applications, but nothing came of it. As historian Milton Lomask explains, academic scientists "were fearful that Federal assistance on a large scale would pave the way to Federal control of their institutions."[2] They got over that fear soon enough.

J.T. Bennett, *The Doomsday Lobby: Hype and Panic from Sputniks, Martians, and Marauding Meteors*, DOI 10.1007/978-1-4419-6685-8_3,
© Springer Science+Business Media, LLC 2010

The contrast with the prewar years was stark. "With the sole exception of aeronautics ... the military played only a marginal role in an interwar political economy of science presided over within the universities by the Rockefeller Foundation and other private philanthropies, and in the corporate sector by the burgeoning laboratories of General Electric, AT&T, and DuPont, and their many smaller imitators," wrote Stuart W. Leslie in *The Cold War and American Science: The Military-Industrial-Academic Complex at MIT and Stanford* (1993). After World War I, scientists went back to mufti; research returned to the private and academic sectors. Yet "For science, World War II marked a more decisive turning point than had World War I."[3] There would be no total demobilization after this war.

As late as 1940, as the historian of science Simon Rottenberg has written, federal spending on science, both basic and applied, was "very small," and what expenditures there were tended to be for agriculture and national defense. All that changed with the war. Within four decades, more than half of all scientific research and development in the United States would be subsidized by the federal government, and about half of that was for military purposes. Rottenberg estimates that by the end of the Cold War, the feds financed "70 per cent [*sic*] of all basic research that is done in the United States and about 50 per cent of all applied research."[4]

The official marriage of state and science in America was, fittingly, a conjunction of the New Deal and World War II. President Franklin D. Roosevelt brought the couple together, though it would be a cold Yankee, an abrasive New England Republican with the unusual name of Vannevar Bush, who drew up the papers and led the pair to the altar.

Vannevar Bush's journey had taken him through several of the realms of scientific research. He had worked on submarine detection in World War I; he had been vice president of MIT; he had done industrial research; he was a founder of the Raytheon Corporation; and he was president of the Carnegie Corporation. At various times, therefore, the military, the university, the industrial lab, and the private foundation had been his home. The federal government was his next stop, as in 1940 Bush was made chairman of the new National Defense Research Committee (NDRC), which joined with the Office of Scientific Research and Development (OSRD), which he also directed. Weapons research and scientific assistance to the war effort were their reasons for being. The "OSRD had outstanding success in bringing science, engineering, and medicine to bear on the conduct of the war," writes historian J. Merton England, but all things must end, even wars, and Bush hated to see an agency that had served the state so well just vanish.[5]

(The energetic Bush also served as chairman of the National Advisory Committee for Aeronautics, which would become NASA in 1958. The guy got around.)

The OSRD was especially helpful in making advances in radar and the detection of airplanes and ships. In wartime the government takes on tasks that it does not perform in peacetime, but often in peacetime there is no reversion to the status quo ante: instead, the ante had been upped, the state has its foot in another door, and it refuses to leave. Such was the case with science. Even as the war wound down, technocrats proposed ways of extending the life of the OSRD, though under a different name — for instance the Office of Science and Technology proposed by Senator Harley Kilgore (D-WV) in 1944.

(The OSRD's biggest contract went to ... MIT, Bush's institution. But none dare call it pork barrel. In fact, MIT "emerged from the war with a staff twice as large as it had been before the war, a budget ... four times as large, and a *research* budget 10 times as large," writes Paul Forman.[6] MIT must be counted among the war's domestic winners. And speaking of winners, not so coincidentally, perhaps, Raytheon, which Vannevar Bush helped to found, was making 88% of its sales to the federal government by 1959.)

As the American historian of science A. Hunter Dupree writes, the OSRD "did not take upon itself the job of making over American science."[7] It oversaw, it supervised, it brought together existing private, academic, industrial, and government institutions for the war effort. Its brainpower came from civilian scientists working in private or university labs under military contracts. Bush saw the OSRD as temporary, to the disappointment of those socialists and statists who thought that it was only the beginning of a revolutionary transformation. Bush put the quietus to the OSRD after the war, but there would be new bureaucracies in its place.

Most scientists working for the OSRD wanted it shut down at war's end. As Dupree writes, "They wanted to go back to the universities if not to their old penury."[8] They had had quite enough of being government employees; now was time to renew the old passions, to follow their own interests, to investigate matters that they wanted to investigate — to be, once more, their own men, not order-takers for a vast bureaucracy.

One man who was even more pleased than most to see it disappear was Frank Jewett, whom we will meet later as the doomed star of this chapter. He said of the OSRD in testimony to Congress in May 1945, "The whole thing is repugnant to the ordinary civilian-life ways of scientists — the restrictions under which they have to operate and the cellular structure. The uniform experience in talking to all of the men who have given a lot of time and effort, to OSRD, is that they want to get out of this thing and get back to their work as soon as possible."[9] Others had different plans.

This phase of our story begins November 17, 1944, with a letter from President Franklin D. Roosevelt to Dr. Bush, then Director of the OSRD, requesting of Dr. Bush a series of recommendations on the relationship between science, war, and government. The central question asked was this: "What can the Government do now and in the future to aid research activities by public and private organizations? The proper roles of public and of private research, and their interrelation, should be carefully considered."[10]

Bush's biographer Z. Paschal Zachary sets the scene. According to Bush, "Roosevelt called me into his office and said, 'What's going to happen to science after the war?' I said, 'It's going to fall flat on its face.' He said, 'What are we going to do about it?' And I told him, 'We better do something damn quick.'"

Action! Take action! Government action! Appoint a committee of experts! Create a federal agency to oversee science! Never mind that science had got on quite well in the first 160 years of the republic without a centralized government agency to guide it.

As Zachary describes the Bush–FDR mindset, "technological and military power were now too intertwined for the nation to return to its prewar neglect of scientific research." The feds would now support basic research, without any specific utility, rather than simply practical applications of science. As for the rabble, "Ordinary citizens must understand that researchers deserved tax dollars even in times of peace."[11]

Dr. Bush was expecting this letter from President Roosevelt asking him to report on how the nation might employ science and scientists in peacetime: indeed, he more or less wrote it. Upon its receipt he set about creating four committees "of distinguished scientists and other scholars" to, in the words of later National Science Foundation director Alan T. Waterman, explain what "the relations of government to science should be."[12] Patron? Antagonist? Indifferent?

What do you think?

FDR's letter tasked Bush with a rather wide portfolio. Fighting disease, "developing scientific talent in American youth," and "aid[ing] research" were among the activities he has to investigate.[13] The four committees Bush set up were a Committee on Science and the Public Welfare, a Medical Advisory Committee, a Committee on Discovery and Development of Scientific Talent, and a Committee on Publication of Scientific Information. The members thereof were high muckamucks in their fields, well-connected and usually acutely aware of their own importance. Each committee transmitted a report to Dr. Bush, the dates of transmittal ranging from January to June 1945.

Science — the Endless Frontier, Bush's synthesis of and elaboration upon the committee reports, was then transmitted to the new president, Harry S. Truman, on July 5, 1945. It became, as Waterman said, "a classic expression of desirable relationships between government and science in the United States."[14] That's one way of putting it. *Science — The Endless Frontier* can also be seen as a rejection of the traditional American policy of laissez-faire and maximum liberty for science and researchers and the official enunciation of a policy of centrally directed scientific research in the service of the United States government.

"The pioneer spirit is still vigorous within this Nation," Bush assured Truman, but now it must be harnessed for the good of all people, and "all people" is defined conterminously with the U.S. government.[15]

The first thing Bush made clear in the report is that there would be no going back. Intimate federal government involvement in science was here to stay. "Science can be effective in the national welfare only as a member of a team, whether the conditions be peace or war," he began. Genius would henceforth be subordinated to the demands of the collective.

President Roosevelt, in sending the initial letter to Bush, was inviting a certain response. Before asking the question he knew the answer. And Bush did not disappoint. "The Government should accept new responsibilities for ..." was the typical beginning of a paragraph, and those responsibilities would include "promoting the flow of scientific knowledge," "the development of scientific talent in our youth," and "the opening of new frontiers."[16] The "lesson is clear," Bush wrote, and the lesson was that if big government could win the war against Germany and Japan, it could surely open new frontiers of scientific knowledge.[17]

Anticipating that a few mossback reactionaries might object that the more unregulated system that produced Thomas Edison, the Lick Observatory, and the University of Chicago should not be so carelessly discarded, Bush declared that such responsibilities "are the proper concern of the Government" and that "this is the modern way to do it."[18]

The report largely ignored the past. After all, now that we knew the "modern way" to do things, what good was musty old history? Science may be an endless frontier but the whole of the past relationship between science and state in America was disposed of in one paragraph, which referenced, of course, the Naval Observatory and the Department of Agriculture and noted proudly that in recent years the number of federal agencies with a scientific connection had reached forty. And yet previous generations had been dangerously indifferent to this important matter. "We have no national policy for science," Bush fretted, nor even a standing congressional committee devoted to it.[19] The lack of national policy or congressional committee had not led to a corresponding lack of science, but this was not the lesson Bush wanted to draw from the fact. If only Congress had established that Department of Science back in the 1880s — who knows how smart and prosperous Americans would have become?

Emergency — in this case war, and the threat of annihilation — was invoked. "Modern war requires the use of the most advanced scientific techniques," Bush declared in the sort of truism to which no one could really object — except, perhaps, the Vietnamese and all the other Third World peoples who were, over the next couple of decades, to fight against the nations of the First World using techniques that Bush would doubtless have considered hopelessly primitive but that were devilishly effective.

He quoted a joint letter from the Secretaries of War and the Navy to the National Academy of Sciences in which the secretaries warned that "war is increasingly total war, in which the armed services must be supplemented by active participation of every element of the civilian population."[20] Like men, science must either enlist or be conscripted. Our enemies engaged in total war; Americans must do no less.

"The major recommendation" of Vannevar Bush in *Science — the Endless Frontier* was that "a 'National Research Foundation' should be established by the Congress to serve as a focal point for the support and encouragement of basic research and education in the sciences and for the development of national science policy."[21] It would take five years from the time of Bush's report for this to come to fruition, and if it — the National Science Foundation (NSF), as it would be called — did not quite meet Bush's lofty description as "an independent agency devoted to the support of scientific research and advanced scientific education," it did change the terms of debate on the subject of American science and its relationship to the federal government. After this debate, the question was no longer *should* there be such a relationship; the only question was how large a chunk of the budget should be devoted to this activity.

Science — The Endless Frontier's description of the National Research Foundation (later NSF) was modest enough: this new agency would be "adapted to supplementing the support of basic research in the colleges, universities, and research institutes, both in medicine and the natural sciences."[22] That word "supplementing" was designed to cool the fevered brows of any old-fangled constitutionalists out there, or any stubborn intransigent scientists who distrusted arbitrary authority.

In addition, Bush stated as fact rather than merely recommended the following: "If the colleges, universities, and research institutes are to meet the rapidly increasing demands of industry and Government for new scientific knowledge, their basic research should be strengthened by use of public funds." To do any less would be a dereliction of duty. True, in the past, industries such as General Electric and AT&T had paid for their own cutting-edge laboratories and research facilities, but that was so 1920s. Modern times demanded public subsidy.

Higher education was about to enjoy an unprecedented expansion due to the G.I. Bill and other inducements to matriculation. In many schools, the hard-science and engineering departments were becoming virtual adjuncts of the War (soon to be Defense) Department. The Bush panel was an early advocate of a closer bond between the university and the state, calling for federally funded science scholarships and fellowships.

Americans were in for a proliferation of bureaus, agencies, and departments. The Bush report urged that a "permanent Science Advisory Board" be created to advise the President and Congress on scientific matters.[23] It was a technocrat's dream: influence on important questions of government policy without any accountability whatsoever, since the advisory board's members would be unelected.

The Bush-proposed National Research Foundation would be divided into five divisions: Natural Sciences, National Defense, Medical Research, Scientific Personnel and Education, and Publications and Scientific Collaboration. Its projected budget would grow from $33.5 million in its first year to $122.5 million in its fifth year. Not all divisions would be created equal. Far and away the smallest share of the outlay ($1 million in the fifth year) was allocated, in Bush's proposed budget, to publications and scientific collaboration — that is, the actual dissemination of scientific knowledge.

The third of Roosevelt's planted questions was the one by which Bush guaranteed his legacy. "What can the government do now and in the future to aid research activities by public and private organizations?" As Bush's biographer G. Paschal Zachary writes in *Endless Frontier: Vannevar Bush, Engineer of the American Century* (1997), "Embedded in this question was the OSRD model of state-funded but privately executed research." In other words, the question was the answer. And the answer was the creation of a federal science program commensurate with the massive stretch of the New Deal.

The establishment of four committees of distinguished establishmentarians to respond to Roosevelt was a neat trick by Bush, "giving the patina of democracy to what was essentially Bush's personal agenda," writes Zachary. Many committee members had OSRD ties. "No one served on a committee who was not a corporate executive, a lawyer, a professor or an administrator of a university, research institute or foundation." This was a self-selected elite class that had grown used to government funding over the last several years. The chances that it would recommend to President Roosevelt that scientific research return to the private sector were less than nil.

This was no mere bureaucratic report with an envisaged shelf life shorter than an Arctic growing season. Vannevar Bush saw the report as "a kind of creation myth, a founding story about the new world conceived by the union of science and government during the war." Scientists had won the war (with a little help from 16 million soldiers, to be sure). Now it was their turn to share in the bounty — or at least be rewarded.

> "While not quite espousing a scientocracy, Bush envisioned a technologically advanced America governed by the masters of science and technology," observed Zachary. Masters need to be fed. And the plain fact was that "Only if fattened on public funds could unfettered scientists truly aid the nation."[24]

The letters of transmittal from the distinguished committee members to Dr. Bush are instructive. Probably the most important committee, that on Science and the Public Welfare, chaired by Dr. Isaiah Bowman, President of Johns Hopkins University, was given the job of determining the proper connection between state and science. This question had vexed American statesmen and scientists since the beginning of the republic, but Dr. Bowman and his sixteen co-committeemen (and they were all men) figured it out quickly. "[T]he Federal Government, by virtue of its charge to provide for the common defense and general welfare, has the responsibility of encouraging and aiding scientific progress," they wrote. The general welfare clause is indeed elastic, but there was no indication in the report that the Bowman committee had paid much if any attention to the constitutional objections to state-supported science that had held sway in the 19th century. War, the New Deal, and modern times had changed everything, it appears. A national government that took "a more active interest in promoting scientific research" would provide "increased security and increased welfare" for the country.[25] Case closed.

The Bowman committee's report did offer a brief gloss on pre-New Deal American science. In the beginning it "was carried on in random, sporadic fashion." How different were those enlightened despotisms across the sea, which may not have protected liberties but they did dole out the dough!

The American laissez-faire philosophy, reported the committee, was "in marked contrast to the principal European countries where, almost without exception, science was directly supported by the governments." At one time this example of American exceptionalism would have been a distinction, but by 1945, it was something to be ashamed of.

So, too, in the Bowman committee's eyes, was the reluctance of the federal government in the 19th century to put its stamp of approval on a national university and other proposals for an expanded governmental role in the promotion of science. Congress, Bowman complained, "turned a deaf ear to all proposals for creating scientific institutions having anything but limited and strictly utilitarian purposes" — this despite "eloquent expressions by scientific men" asking for aid.

Bowman and company denied that dissent even existed today over the basic constitutional question. "As far as the committee can determine, there is no major dissent from the view that" subsidizing research laboratories and universities falls "within the proper function of government." It depends, as we shall see, on what one means by "major dissent."

As for direct subsidies to universities, the alleged crying need for such provided all the justification necessary. Yes, Bowman's committee admitted, there is a "fear that Federal aid might lead to centralized control," but since "universities clearly stand in need of increased financial support," there is really no use arguing the point. The universities want money; the universities must have money. The need, we are told, is "imperative."[26] (The Committee on Discovery and Development of Scientific Talent, which called for the establishment of undergraduate scholarships and graduate fellowships for the study of science, was stacked with Ivy Leaguers. Of the six members listing academic affiliations, three were Harvard-connected — including the school president, James B. Conant; one was from Brown; one Princeton; and the outlier was Dr. T.R. McConnell, Dean of the College of Science, Literature and the Arts of the University of Minnesota.)

Those members of the Bowman committee who had doubts about the wisdom of federal subsidy of academic research had as a spokesman — of a kind — Reverend J. Hugh O'Donnell, president of Notre Dame, who explained, "This is a case of venial sin. It can be forgiven in view of the good which it may do."[27] The Jesuits call this kind of ethical calculation casuistry.

The mid-1940s were the peak years in the reputation of the state. Private enterprises were viewed with suspicion; government, which had just won a war against tyrannical fascism and its allies, was lionized. There was nothing it couldn't do. So in discussing the wisdom of federal aid to universities,

Dr. Bowman cast a jaundiced eye on *private* aid, especially from industries. He remarked that "the freedom of the university scientist may be decreased by the introduction of some degree of commercial control." It was *commercial* control, not governmental control, that was to be feared. In fact, absent a massive infusion of federal cash, "there is danger that an undue number of trained individuals may go into industry, stripping the universities" of competent scholars and researchers. It was the federal government's solemn obligation, its duty, therefore, to "maintain a favorable competitive position for universities relative to industry."[28] That "favorable competitive position" was to be maintained by the taxpayers.

Dr. Bowman's committee was the one that produced the recommendation for a National Research Foundation to distribute funds to scientific research institutions, "initiate and finance" projects in government agencies, "[e]stablish scholarships and fellowships in the natural sciences," and disseminate "scientific and technical information."[29] (In an aside, it expressed the hope that the provision of federal funds for research in the natural sciences would free up additional monies for the social sciences.) This NRF was also to "provide adequate facilities for advanced research" and fund postdoctoral research as well.

So far, so good, at least as far as the government-science establishment is concerned. But the devil, at least to the Truman administration, was in the details. The democratic process was useful for opening the pipeline between the feds and scientists, but Dr. Bowman was not about to counsel democratic oversight of the pipeline. He and his committee proposed "a largely autonomous board with a staff of men trained in science" to run this National Research Foundation.[30] In other words, experts insulated from politics, from democracy.

In the words of biographer G. Paschal Zachary, "Bush believed Americans should freely give public-spirited experts ultimate authority over the nation's security." If that weren't enough, "Bush worshipped at the altar of rational planning."[31]

The National Research Foundation's trustees were to be "eminent men of science who are cognizant of the needs of science, and experienced in administration." This seemed reasonable, at least to a man with zero experience in practical politics. Dr. Bowman recommended a 15-member panel serving five-year terms. These eminent trustees were to be appointed by the President of the United States "from a panel nominated by the National Academy of Sciences," and in turn they were to appoint the executive director, who would serve in a full-time capacity.[32] Therein lay the rub, or the beginning of the five-year rub.

For what caused the five-year delay between the appearance of the Bush report and the signing of the legislation creating the National Science Foundation was a struggle over power. Not whether or not the power would exist: the Bush forces and the Truman administration had no quarrel with that. Rather, the tussle was over who would control the NSF — politicians or scientists. Guess who won?

Vannevar Bush, in adapting the Bowman report, proposed a nine-member board for this National Research Foundation. The members, who were, he wrote with a straight face, to be "not representative of any special interest," would be appointed by the President to four-year terms.[33] As in Dr. Bowman's recommendation, board members would elect their own chairman. So while the President could shape the board to his liking, the board members themselves would have the final say over the chairmanship. The scientists thought this just fine: why, they thought, should a grubby ward-heeling politician out of the Kansas City Democratic machine appoint the Grand Pooh-Bah of American Science? (Congress had been cut completely out of this loop, which was not surprising, given the incredible growth of executive power over the previous dozen years.)

The Bush plan had been anticipated by the somewhat dimwitted Democratic Senator Harley Kilgore of West Virginia, who once confessed to "utter, absolute ignorance" of science — a confession that did not keep him from becoming a major legislative player in science debates. Senator Kilgore proposed during the latter part of the war the establishment of a National Science Foundation, which would lead to an unprecedented centralization of scientific research in the service of "economic security."[34] The name "National Science Foundation" was the gift of Senator Kilgore, whose chief staffer on the matter was physicist Herbert Schimmel on his Subcommittee on War Mobilization. The problem was, Kilgore and Schimmel wanted the U.S. government in a kind of permanent state of mobilization, war or not.

In 1942, Senator Kilgore had proposed establishment of an Office of Scientific and Technological Mobilization (S. 2721) in order "To mobilize for maximum war effort the full powers of our technologically-trained manhood."[35] This sounded like an ominous threat, totalitarian in flavor, as the office was to have the power "to draft all such personnel and facilities failing to submit or to accept plans for immediate conversion of their efforts to work deemed more essential by the Office of Technological Mobilization." In plain language, Kilgore sought to make all scientific personnel, even those engaged in fully private ventures, servants of the state, subject to direction by a vast new bureaucracy. This bureaucracy was also to fund research both basic and applied, and provide grants and loans to students.

The redoubtable Frank Jewett denounced the Kilgore proposal as a plan to make scientists "the intellectual slaves of the State." The National Association of Manufacturers chimed in that Kilgore was trying to "socialize" science.[36] But that language was sounding very 19th century in the 1940s.

Supportive of Kilgore's mobilization bill was E. Waldemar Kaempffert, science editor of the *New York Times*, who told the Subcommittee on War Mobilization, "We have followed in scientific and industrial research what the economists call the laissez-faire policy which is now outmoded for economics but which still prevails in research." Kaempffert said that as a result of the world's failure to centrally prescribe research, "progress has been made in a haphazard way ... with the exception of Soviet Russia," where Stalin the Wise was unafraid to use the necessary coercion. Kaempffert's testimony makes for shocking reading today. He unblushingly praised one of the greatest mass murdering regimes in human history, saying that "the only government in the world, I regret to state, that has used science, or intended to use science to secure social security, social happiness and contentment, is Russia." Joseph Stalin: bringer of "social happiness"! Who knew?

Senator Kilgore led Mr. Kaempffert of the *Times* into a paean to Soviet science. "The Soviet Academy of Sciences, which is the equivalent of our National Academy of Sciences, is an integral part of the Government," said Kaempffert, "and as much so as our Department of Agriculture or Department of Commerce. It plans the research activities of the entire country, and those plans extend right down into every shop. The result was that Russia found it least difficult of all the nations to turn over from peace to war." This is undoubtedly true. A government that wages war against its subjects every day will find it easier to wage war against another nation than will a government that respects the liberties of its citizens.

Trading praises for total state power over "every shop" in a vast land, Kilgore and Kaempffert offered a truly chilling portrait of the potential for tyranny that existed even in the capital of the Free World. Kaempffert remarked that the lesson of the Soviet Union was "that technological progress can be made only by competent planning, direction and organization," which he was "glad to see" in Kilgore's bill. Yes, those Soviet five-year plans were models of farsightedness.

These proposals were no mere wartime expedients. As the *Times* editor said, "I hope it will not be just a wartime institution, but that it will be a permanent institution, because, why should we destroy this elaborate apparatus which is bound to produce much good, and then let ourselves drift along aimlessly as we have in the past?"[37]

"Drift along aimlessly": That was how these men regarded liberty.

Waldemar Kaempffert elaborated on his dreams of centralized state-directed science in an October 1943 article for the *American Mercury*.

He praised Senator Kilgore for proposing an Office of Technological Mobilization, which "would marshal the country's scientists and engineers in a vast research organization to solve the pressing war problems of the armed forces and industry." This would help solve what the *Times* propagandist viewed as a pressing problem in scientific research: "There is no overall direction and control."

He contrasted this comparatively anarchic system with those of the Nazis and the Soviets: "It may be assumed that Germans exercise some such control over scientific and technical knowledge. It is certain that the Russians do. In fact, Soviet Russia is the one country that mobilized science not only for war but for peace."

It might seem a bit tacky to be praising Hitler's Nazi Germany and Stalin's Soviet Union in 1943, when the crimes against humanity of both those nations were fairly well established, but Kaempffert isn't holding back. He concedes of his subjects: "Totalitarianism? Of course. But we are waging a totalitarian war." In other words, to defeat a totalitarian power (or, in the case of the USSR, an ally), one must become totalitarian. This would seem to subvert the very purpose of the war, but then to Waldemar Kaempffert, liberty is nowhere on the radar screen. He sneers at the supposed "sanctity of free enterprise." He ridicules the idea that "The hired scientist must be as free as a bird." For to the Kaempfferts, freedom is an illusion. We all live in cages. The only question is who is to be master of the birds.

Perhaps sensing that he has gone too far, Kaempffert protests that "This is no plea for collectivism." Yet this defense is belied by every other sentence in the article. His central contention is that "government should undertake research on a new scale for the common good, so that science will no longer drift aimlessly, as it has drifted in the past. Laissez-faire has been abandoned as an economic principle; it should also be abandoned, at least as a matter of government policy, in science." Newton, Pasteur, Einstein: think how much those aimless drifters of the past might have accomplished if they had government bureaucrats to give them direction!

The Kilgore plan is only a start, says Kaempffert. It "hardly goes far enough," especially since the West Virginia senator provided it with only a ten-year lifespan. "There is every reason why it should be a perpetual plan," says Kaempffert. After all, "We need an all-embracing plan, organization and competent direction" if American science is to thrive.[38]

That such totalitarian fever dreams could be taken seriously shows just how low an estate liberty enjoyed in the debates of the 1940s. The shadows of tyranny were upon the world, darkening even the side of liberation.

Harry Grundfest, president of the American Association of Scientific Workers who would spend much of his career as a professor of neurology at Columbia University, was another proponent of a vast expansion of the state in the area of scientific research. Grundfest was hauled before the Senate Internal Security Subcommittee in 1953 at the height of the McCarthy scare and refused to tell the subcommittee whether or not he had been a member of the Communist Party. Whether or not Grundfest was a communist, the AASW was born in the mid-1930s in a milieu in which, according to the Temple University library, repository of the AASW papers, "some scientists turned their eyes to the Soviet Union, where the five-year plans were in full swing and fully-funded science was being hailed as the state religion."[39] Socialists and communists would see the war as a once-in-a-lifetime opportunity to subordinate, perhaps for the long term, science and scientists to the urgent needs of the central state.

Even in the darkness, one could hear voices of dissent. In June 1943, in the pages of *Electrical World*, L.A. Hawkins, executive engineer at the General Electric Research Laboratory, an oasis of private industry, blasted Kilgore's plan for being based on "revolutionary totalitarian regimentation" of scientists — not only those whose work is directly tied to military research but also geologists, biologists, astronomers, and even archeologists. Indeed, all college graduates or those with at least "six months training or employment in any scientific or technical vocation" would have been drafted into Kilgore's civilian army, whose unelected board would have the power and the funds to buy corporations or set up new government-controlled firms that would perform scientific and technical research.

This was an almost mind-boggling attempt to nationalize science under the cover of war, and to make it worse Kilgore's bureaucracy would "extend indefinitely into peacetime." One suspects that in the view of Kilgore and his allies, there would be no more peacetimes, just endless wars, whether against real or metaphorical enemies — against Russians or "poverty" or "ignorance" or "intolerance." So, the scientific and technical conscripts would be more or less permanent draftees, their condition defined in military terms so as to avoid running afoul of the Thirteenth Amendment. (Who, in 1865, would have thought that slavery might make a comeback within a century?)

L.A. Hawkins ended with a rhetorical gust the vehemence of which had seldom if ever been experienced in the usually un-charged pages of *Electrical World*: "It is not enough, it seems, that the pocketbooks of the industrious and thrifty should be picked to support incompetent and idle individuals, but their brains are to be rifled to support the inefficient industries."[40] This was one engineer who intended to fight for his liberty.

Hawkins's was a minority voice, at least at elite levels. Many liberals had fallen in love with statism, seeing the way that the government had mobilized millions of Americans for the war and wondering why such mobilization would not work in peacetime as well. As the war wound down, *The New Republic* editorialized that "this has been a war of the scientists, more than any other in history. The mobilization of scientific ability in the enemy countries has been discouragingly good, and in some ways better than our own, deeply imbued as we have been with the notion of scientific research as primarily an offshoot of private business, and designed to make money."

Again, we hear the admiration for the Nazi-Soviet way of doing things, or at least doing big science, and contempt for liberty and private enterprise, which were thought to be hopelessly inefficient when compared to Soviet five-year-plans and Nazi edicts.

Once the war is over, *The New Republic* worries, "There is a danger that, with our foolish adherence to so-called free enterprise, we shall simply appropriate funds and wait for somebody to come along and ask for them. On the contrary, research needs to be coordinated carefully and the projects should be selected in terms of our national necessities, and not the accidental interests of various scientific groups. We do not see how this main job can be done by anything less than government itself."[41]

The worry, therefore, is not over the unprecedented mobilization of men and wealth involved in the Second World War. Instead, the worry is that this conscription of men and wealth might end, or at least be seriously cut back. Demobilization is to be feared, according to the liberals of *The New Republic*. The nation on a war footing is so much easier to rule, its resources being right at hand, than is the nation at peace. As the Bowman Committee reported to Vannevar Bush, "During the war, Government expanded its research budget from $69,000,000 in 1940 to $720,000,000 in 1944"[42] — more than a tenfold increase. War is the engine of government, and that includes government science.

Assistant Attorney General Thurman Arnold had expressed concern in 1943 that industrial labs were somehow "monopolizing" scientific knowledge. But, historian Daniel J. Kevles points out that "Arnold was on somewhat shaky ground when he claimed that big business held a commanding corner on scientific research in the United States. ... Actually, industrial corporations provided only a small fraction of academic research funding, and for the most part professors of science and engineering studied what they wanted and published their results in the professional journals."[43]

Yet even so, Arnold and the New Deal liberals were discomforted. Why should the market and the private academic sector determine how scientists spent their time? Shouldn't these eggheads been put to work meeting pressing national needs, those needs to be determined by high-level bureaucrats in Washington? Who did they think they were, anyway — free men and women? A war was on!

In 1945, Senator Kilgore issued a report, from his Senate Subcommittee on War Mobilization, urging "an increase, above the prewar level, in the Government's support of research and development activities in fields that are predominantly in the public interests, notably national defense, health and medical care, and the basic sciences." It called for a National Science Foundation, with a director to be chosen by the President with the approval of the Senate, that would not do its own research but fund private and public research organizations. Kilgore would also empower the foundation "to grant fellowships and scholarships in various fields of science."

Controversially, inventions or discoveries from any research project that received full or even partial federal funding would "become the property of the United States."[44] Senator Kilgore argued, not wholly illogically, that patents that resulted from federal grants should not benefit a private person or interest but rather be in the public domain. Vannevar Bush would respond that such a policy would greatly reduce incentives for new research. He carried the day, though it took him five years to do it.

Foes of subsidized research included California Institute of Technology President Robert Millikan, who warned of "collectivism"; and Frank Jewett, then president of the National Academy of Sciences, who urged Bush in a June 5, 1945, letter to try "revivifying that fruitful stream [of private funds] before plunging into the uncertain waters of the Federal tax pool."[45]

On July 19, 1945, *Science — The Endless Frontier* was released. It was a sign of the times that very few critics scored Bush for going too far. Central direction of science had become a consensus opinion. (Although Truman's director of the Bureau of the Budget, Harold Smith, did ask if the "endless frontier" would lead to "endless expenditure."[46]) Rep. Wilbur Mills (D-AR) and Senator Warren Magnuson (D-WA) introduced Bush's proposal as legislation, but it stalled in 1945. President Truman would not endorse *The Endless*

Frontier, largely because the director of the proposed National Research Foundation was to be selected by the board rather than the president.

In the Bush–Kilgore debate — or, more accurately, difference-splitting — of 1945–50, the Bushites did not favor the inclusion of the social sciences under an NSF, while the Kilgorians did. Nor did Bush want the kind of radical change in patent policy favored by Kilgore. And, Kilgore the executive-power Democrat wanted the President to control the NSF, while Bush the technocrat favored a nonpartisan board. Kilgore wanted geographical distribution of funds, while Bush favored channeling them to elite institutions. Yet despite the differences between Bush and Senator Kilgore, "At bottom, they both wanted the government to pay for independent research, with almost no strings attached."

In 1946, something of a hybrid of the Bush and Kilgore plans was introduced. Frank Jewett called it "the perfect vehicle to socialize and nationalize a large independent section of our economy." Republican House members charged that it was a "link in the chain to bind us into the totalitarian society of the planned state."[47]

In July 1947, Congress passed a Bush-tinctured National Science Foundation bill, but Truman vetoed it on August 6, 1947, because the board, and not he, selected the NSF chief. Truman had no basic disagreement with a central fact of postwar science: that is, that the federal government had become the primary dispenser of R&D funds in the United States. It was all about executive, that is, presidential, power, for him. The problem was the bill as passed by Congress created an NSF that "would be divorced from control by the people to an extent that implies a distinct lack of faith in democratic processes." What Truman meant by "the people" was "the president." By making the director answerable to the 24 members of the foundation rather than the president, the legislation led to a "lack of accountability."[48] Back to the drawing board it was.

As in all cases when government dispenses funds, the distribution formulae were a matter of heated dispute. This is the question that underlays many legislative debates: Who gets the loot?

In support of a wide geographical distribution of NSF funds, Clarence A. Mills, a Westerner, wrote in *Science* in 1948: "There exists no evidence that native intelligence is better in one part of the United States than in another.... The time has arrived when the West should shake off the stunting dominance of the northeastern seaboard in scientific matters, insisting on autonomy and a just share of public funds for its scientific development. So long as the rich eastern institutions secure the major part of funds disbursed, western institutions will perforce remain relatively pauperized and their most promising young scientists drift eastward, where working facilities are more propitious."[49]

Harley Kilgore would have allocated money on a state-by-state basis; Bush called for individual contracts with the new agency. Then again, his experience within elite institutions led him to believe that the federal government should be subsidizing the MIT professor and not the professor at Montana State. Ultimately, the result of the NSF and related programs was "a relatively decentralized federal science establishment, although one dominated by national security interests, in which academic research support comes in the form of project grants awarded to individual university researchers," as James D. Savage wrote in *Funding Science in America: Congress, Universities, and the Politics of the Academic Pork Barrel* (1999).[50]

The most important congressional hearings in the history of the debate over a federal science foundation were held in the Senate in October 1945. A parade of witnesses testified to the pressing need for such a program. Of the one hundred scientists, bureaucrats, and administration figures who testified in the fall 1945 hearings on the NSF proposals, there was but a single naysayer: Frank B. Jewett.

Jewett brought to the table a resume that was almost unequalled in American science. Forty-five years of both fundamental and applied science were in his background: he had been, among other things, vice president of American Telephone & Telegraph, head of Bell Telephone Laboratories, president of the National Academy of Sciences from 1939–1947, and a key member of the National Defense Research Committee of the Office of Scientific Research and Development. He was not a crank who wandered in off the street, or a feverish ideologue spouting an untested philosophy. His background was old-fashioned American: his ancestors had settled in New England in the 1630s, and his parents were among the earliest residents of Pasadena, California, where he was born. He received his A.B. from Throop Institute, later California Institute of Technology, when he was just 19, and his doctorate from the University of Chicago. He taught physics at MIT before embarking on his career in industrial research. At AT&T, his first post-academic posting, he displayed "the keen analytical ability, the refusal to be blinded by superficial obscurity, and the perception of essentials that were characteristic of him throughout his life," writes Oliver E. Buckley in the biographical memoir of Jewett he prepared for the National Academy of Sciences.[51] Beginning as a transmission engineer, Jewett, by dint of hard work and his native genius, rose to the vice presidency in two decades. Seemingly universally liked and respected, he was that unusual scientist who displays a kind human touch in his everyday doings. And he saw the growth of government science as, in some ways, contrary to the best aspects of the American nature and the American state.

In the postmortem assessment of historian John C. Burnham, Jewett was "a conservative who … was conscious also of unfavorable tendencies engendered by the role of science in the war and the period just after…. Testing his prescience by later events shows him to have had impressive insights into his own times."[52]

Milton Lomask, the historian of the National Science Foundation, writes somewhat patronizingly that "A faint nostalgia wafted from Dr. Jewett's remarks." How quaint! "He wanted scientists to continue as in the past to get the bulk of their financial help from private enterprise and private foundations…. His advice to Congress was that instead of erecting another Federal bureau, it change the tax laws so as to encourage greater support of science from private sources."

"Jewett's testimony was the voice of the 19th century talking to the 20th, and probably the splendid old electrical engineer was as aware as anyone present of how deaf every century is to its predecessor," observes Lomask. Heroically, Jewett "would persevere in his crusade to save his fellow scientists from what he regarded as the imprisoning embrace of Federal largesse."[53]

This is not a disrespectful characterization of Jewett's remarks, but they are not perhaps quite as nostalgia-tinted as the NSF historian would suggest. Let us listen for echoes of classic American themes in Jewett's statement and response to questions of October 18, 1945, and consider whether he may not have been quite the fossil he is made out to be.

Only three senators were present, all of them Democrats: Senators J. William Fulbright (AR), Warren Magnuson (WA), and of course Harley Kilgore (WV).

He was "in opposition to much that has been said," Jewett admitted, but he intended to have his say, for "they are my honest opinions and as such you are entitled to hear them."

Jewett did not oppose a role for the federal government in wartime research. To the contrary, he called OSRD "the greatest industrial research organization the world has ever known." In applying the latest scientific knowledge to problems of weaponry and war, the office was both profligate and productive. "Measured by any standards of active war, where time and results warrant any expenditure of money and wasted effort, the results astoundingly and fully justified the undertaking," Dr. Jewett told the three senators and the audience in the hearing room. But "Measured by any standards which a peacetime economy could possibly support, it was wasteful in the extreme."

Jewett thus cut immediately to the core of the problem with the proposals by Vannevar Bush and Senator Kilgore. They were predicated on the permanent existence of war and a wartime economy. They did not envision peace. They assumed that the extraordinary circumstances of war had produced great things in American science and technology, and they wished to keep science and technology on a war footing.

But this could not be done without conscripting science and scientists. During the war, said Jewett, "men of great ability gave their full-time service gratis — a thing they neither could nor would do in peacetime."

To place American science on a more or less permanent wartime emergency footing would be "a radical departure from the normal American way, which during the past 150 years has taken American science and technology to the highest pinnacle man has ever achieved." That this summit was achieved by free men whose inventiveness and genius was nurtured by liberty and not state coercion was a fact ignored by the would-be architects of Big Science.

"It may be," said Jewett, "that conditions have so changed permanently that we must abandon the old way which depended on the voluntary action of free men operating in the thousand and one ways that men choose and turn to the State for a large part of the support of science, through a politically controlled agency. As yet, I, for one, am completely unconvinced either as to the necessity or desirability of the change so far as fundamental science is concerned."

Across the board — in technology, physics, agriculture, and other branches of science — the case for state subsidy was weak. And it carried with it insidious dangers. Jewett had learned the lesson that "Every direct or indirect subvention by Government is not only coupled inevitably with bureaucratic types of control, but likewise with political control and with the urge to create pressure groups seeking to advance special interests."

Why throw the old system, which had realized such exceptional achievements, out the window? What was wrong, Jewett wondered, with the methods of financing which led America to become the world leader in, for instance, astronomy, within the span of a few decades?

"Men of substance here in America have always been liberal in aid of fundamental science, and I have no reason to think that any present diminution of personal gifts is necessarily permanent." In addition to industry labs of the sort which Jewett had managed so successfully, new nonprofit research organizations held out great promise. And that promise was a function of the freedom they enjoyed — a freedom not possible within the agencies dreamt of by Kilgore and Bush.

"Calling an agency of Government a Research Foundation does not make it a research foundation in the sense that we have come to know them," said Jewett to the senators. "Such private foundations, established by outright gift to self-perpetuating boards of trustees, can act with a freedom impossible in an agency which derives its money annually through the intricate channels of Government appropriation." The government foundation, no matter how pure the intentions of its creators, will inevitably be "politically controlled," and "because it is a politically controlled agency subject to many indirect pressures, it will be impossible to turn down many second-rate applications."

Under questioning by Senator Fulbright, Chairman of the Subcommittee on War Mobilization, Dr. Jewett insisted that "the private thing is far freer to operate efficiently than is anything which necessarily has to be subject to the multifarious political influences." Wouldn't relying on private monies be less democratic than using taxpayer money for research, asked Senator Fulbright, apparently affronted by Jewett's implied criticism of political control.

> "[Y]ou infer that politics is wrong," said Senator Fulbright, incredulously. "You find me a single big corporation that doesn't have a lot of internal politics and I will go out and find you some vice president's son-in-law who is drawing a big salary, who isn't competent."

Touché? Or not?

Jewett fired right back. "All right. Take the one I was connected with. Go ahead and find him. Go ahead and find him. You picked a bad one, I think, because one of the things I found when I went into the telephonic company more than 40 years ago was that the proposition of having your relatives or sons or friends or what not in there was so looked down upon that men couldn't even have their own sons working in the company."

Fulbright replied, not entirely convincingly, that AT&T was better than mere grungy business because it was a regulated monopoly, but Jewett would not let him off the hook that easily. "No, it was not when I went in it; there was not any supervision by the Government at that time."

Jewett conceded nothing of the superiority of government (or government-supervised monopolies) over private enterprise. In fact, he saw such enterprise as the hope of American science. Playing a favorite theme which he had played and was to play in other public statements, Jewett said that by removing "tax laws" and other "dams which have diminished the flow of money" into private and non-profit foundations, individuals and industry will provide more than enough financing for basic research and fundamental science to flourish. In such circumstances, "there wouldn't be any necessity for taking taxpayers' money and trying to set up under what must of necessity be a politically controlled thing, subject to a million and one controls which the private fellow doesn't have, the private institution doesn't have."[54]

Another piece of testimony by Jewett is worth quoting at length, for his is the path that was offered but not taken:

> I think I would like to start by saying that my approach to this question of a Federal organization to dispense money in this field is the approach from a background of nearly 50 years of experience in the fields of science, both fundamental and applied, and engineering; and, of course, almost that amount of time in the observation of the way in which we here in America have, in the past, conducted our various institutions.... .

> The reason that has been given for the diminution of the amount of money which would be available for science and other things is the operation of the tax laws. Those are man-created things. If then, that being the case, Congress should decide that science should get more money than it is likely to get under the existing tax laws and should decide that it is more profitable to have that money flow in the traditional way, the process of directing a flow of money by voluntary giving is quite simple, viz, by changing the tax laws to stimulate increased voluntary giving.

> Now, over the years, if one looks at the history of giving in this country, one finds that at one period of time men and women were particularly interested in religion and churches, and the major part of the giving was in support of these, and at other times it was education, or medicine, or hospitals. More currently it has been in the fields of science. Always the pattern has been a gradually changing pattern, as men's interests changed. It has always been voluntary, however; and, in the main, it has resulted, I think, in a reasonable distribution of the best brains of the country into the several fields in which a complex civilization like ours has to be fructified with able people if the thing is going to function. It would be a calamity, in my judgment, to have one sector, whether it be law, or medicine, or engineering, or science, or what have you, elevated by law way above the rest. Our economy simply would not work efficiently.... .

> So on two scores I am skeptical: First, as to the premise on which the necessity for this foundation scheme is based, and, in the second place, I am very doubtful as to whether the people of the United States, for the money which is spent in this particular sector, will get as much value of their dollars as they would get were they in a position to make the expenditures directly in the traditional way.

> Further than that, it seems to me that if you set up a Federal corporation and furnish it with funds to spend in a particular sector of our economy, you are not only running the danger of overstimulating but you are certainly inviting the formation of pressure groups in the other sectors — the social sciences or a thousand

and one other things — who feel that they have a valid claim of special treatment, and you are going to be confronted, in my judgment, in the not very distant future, with urgent claims to set up corresponding corporations to fructify other fields. [He did like that strange word "fructify."]

So from my own personal experience, and believing as I do that if you are sincerely interested and feel that science should be pushed forward faster than it is likely to go under present conditions and you want to get the most that you can out of it, I am quite certain that if you are willing to change the tax laws so as to put a slight premium on eleemosynary giving to those who give to science rather than to churches or hospitals or what not you will have more money through private channels than you can possibly justify through a channel of this kind. If you do it that way, you do not disturb any of the existing experiences — any of the existing channels of giving which have made this country great. At the same time Congress would not be subjecting itself to the political pressure which I think would be inevitable if a national science foundation is created.[55]

That was the voice of the old free enterprise pre-New Deal America speaking. Science, said Jewett, had no special claim on the public purse, and besides, more could be raised — and in a more just manner — by appeals to private interests than by lobbying the federal government for handouts.

Jewett's was a lonely voice in the hearings on the Bush and Kilgore bills. If anything, voices clamored for *greater* politicization of science. For instance, Rep. Maury Maverick (D-TX) castigated Dr. Isaiah Bowman, president of Johns Hopkins University, who had said that he worried about "political management" of science. Bowman did not oppose federal subsidy of science, but he wanted the important decisions to be made by scientists, not bureaucrats. Maverick was steamed and told Bowman he "need not be so smug."[56] The people, after all, had been wise enough to elect Maury Maverick to Congress. And the Maury Mavericks had just won a war, unlike the geek hirelings of private industry and academia.

Maverick's rebuke to Bowman was occasioned by this rather haughty pronouncement by the doctor: "Do not open doors for untrained and worse than worthless employees who may creep into positions of control and attempt to pass themselves off as wise administrators who understand better than so-called fuzzy scientists how the job should be done." Ouch! (Dr. Bowman later chaired the Committee Supporting the Bush Report, which was created to lobby for an NSF controlled by scientists and not, as in the Kilgore bill, the president.)

Rep. Wilbur Mills (D-AR), later famous as the chairman of the House Ways and Means Committee whose drunken nighttime dalliances with a stripper named Fannie Fox ended his career, introduced on July 19, 1945, a Bush-based bill to create a National Research Foundation. Senator Warren Magnuson (D-WA) introduced the Senate companion. Kilgore introduced his own bill, and the battle was joined. Yet, "Actually, the two bills were much alike," the primary difference being over patent policy and who, ultimately, would run the show. (Bush somewhat patronizingly said of Kilgore that while he was "honestly trying to get at the root of matters," "some of the people about him steer him into strange paths."[57]) On the fundamental underlying question — should the federal government subsidize scientific research? — the principal players were in agreement, leaving the lonely job of dissent to Frank Jewett and a few right-wing Republicans in the Congress.

Squabbles over who was to hold the real power over the NSF — the scientific clerisy or the president — delayed passage of various NSF bills for the next four years, as did the opposition of influential House Republican James Wadsworth, an aristocratic New Yorker who, "influenced" by Frank Jewett, opposed the legislation "on economy ground[s]," and used his position in the House Rules Committee to prevent the discharge of the bill.[58]

The indefatigable Jewett continued what columnist Marquis Childs unsympathetically called his "one-man crusade" against the NSF, persuading Senator Hugh Scott (R-PA) to offer legislation embodying his belief in tax credits for private scientific patronage.

Watching the feuding between partisans of the Bush and Kilgore approaches, Jewett wrote in August 1946 that "Altogether I fear that scientists haven't raised themselves in the opinion of thoughtful men by their performances these past few months." A scramble for public funds is seldom an edifying spectacle.

The Republican minority on the Senate Committee on Military Affairs took a stand against Kilgore–Magnuson in 1946, with a majority of its members calling S. 1850 "a clear exposition of the philosophy of centralization and control of science with its attendant bureaucratic autocracy." The bill would bring a "large sector of our national economy ... under the centralization, control, and supervision of Washington" and was "a link in the chain to bind us into the totalitarian society of the planned state."[59]

Senator Raymond E. Willis (R-IN) was a Jewett-allied skeptic, as were Senators Robert Taft (R-OH) and Kenneth D. McKellar (D-TN). Another Democrat speaking in Jeffersonian accents was former Democratic Rep. Fritz Lanham of Texas, who told a House subcommittee in hearings on NSF legislation that "The American pre-eminence we have attained has not been based upon the centralization of arbitrary power in government. That's a doctrine of certain foreign regimes, and it is becoming increasingly evident that they wish us to adopt it to their own totalitarian advantage. Our progress has come from that rugged individual upon whose freedom of action our governmental philosophy has been predicated, an individual laboring under the incentives of our Constitution and laws to give full expression to his God-given ingenuity in creative accomplishments for our progress.... From the cellars and garrets of these humble, but independent, folk have emanated the forces of American greatness. Let us be sure to remember that."[60]

Lanham's view was considered ridiculously anachronistic. He belonged to the age of leeches, of phrenology, of alchemists claiming to transmute dross into gold. Or that is how "progressives" viewed him.

Yet the fear of Jewett, Lanham, et al. of centralized control of science by an expanding state was no Chicken Little fantasy. Big government was a god to Senator Kilgore and his allies. As former New Deal official Morris L. Cooke testified, "We favor the coordination of all Government-financed research whether in private or public institutions by a National Science Foundation in order to further efficient planning and the most comprehensive use of facilities."

Frank Jewett's letter to Vannevar Bush of March 29, 1946, sounded a note of desperation: "Senator Kilgore's left-wing staff is already making plans to take over and 'guide' the work of the Foundation. They have launched a study to 'plan a science program for the United States.'"

"If the scientists don't awaken to their danger, this group will be able to have their way. S. 1850 [the joint Kilgore–Magnuson "compromise" bill] is the perfect way to socialize and nationalize a large and independent section of our economy."[61]

(Jewett had a genuine concern for liberties that he believed were slipping away. He opposed FBI investigation of Atomic Energy Commission fellowship recipients because, as he told Bush in a letter of June 3, 1949, "If Congress sets such a pattern it will also inevitably crop up in Science Foundation fellowship awards if such a Foundation as you favor is established and then we will be in the race started by the Nazis and Moscow."[62] Many liberals saw nothing wrong with loyalty oaths, especially under liberal presidents.)

Still, some scientists resisted the notion that they were henceforth to be servants of the state whose primary function would be furthering the military aims of their governments. For instance, John R. Baker, biologist, anthropologist, and zoologist, was an Oxford professor who with chemist–philosopher Michael Polanyi had in 1940 founded the Society for Freedom in Science, which defended the liberties of scientists against the demands of the state.

"[I]t would be futile to try to confine [the good researcher] within the rigid boundaries of a central plan for the advancement of science," said Baker. Praising the inspired amateur, the solitary genius, the researcher who follows his own path, Baker argued that "scientific discovery is often prompted by extraordinary and unpredictable circumstances apart from the material needs of man." It is not easily or usefully subordinated to the collective. As Albert Einsten said, "I am a horse for a single harness, not cut out for tandem or team-work."[63]

Baker quoted Hungarian physiologist Albert Szent-Gyorgyi, Nobel Prize winner in medicine, who scoffed at those who suggested that he set out to "discover something useful." He was following his own lights, insisted the Nobelist, who said that any great discovery "was spurred on by an inner passion and impulse."[64] What government bureau could ever understand inner passion and impulse?

Baker's was a voice for classical liberalism in an era of totalitarianism; he believed that science can be an end in itself — it need not necessarily serve some material purpose. It cannot be directed by a central command; the minds of scientists are not mere instruments of political authority, to be channeled down socially approved lanes. As Szent-Gyorgyi avowed, "The real scientist … is ready to bear privation and, if need be, starvation rather than let anyone dictate to him which direction his work must take." He must be free to make his own inquiries, to set his own research agenda, to seek out truth in his own way. And these imperatives are just not possible when the government is financing and overseeing his work. (Baker hastens to add that, of course, the scientist's freedom of inquiry does not constitute a claim upon the purse of others. "A scientist who wants to work at the tree-top fauna of the Brazilian forest can generally be free to do so only if he can persuade other people that his study is likely to be fruitful; then and then only will funds be forthcoming." But, he adds, "An enormous amount of scientific work can be done … with little expenditure."[65])

Let us take a step back from the NSF's ultimate enactment, for the private approach that was shunned in the late 1940s had a recent antecedent, however compromised it may have been: the National Research Fund, a brain-child of George Ellery Hale and Herbert Hoover.

By the 1920s, General Electric and American Telephone and Telegraph (AT&T) had established well-endowed and impressively staffed research laboratories. As Joseph P. Martino wrote, "one of the lessons of the inter-war years is that if industry funds research adequately, a significant amount of basic research will be carried out even though the bulk of the research is applied."[66]

The seeds of the National Research Fund idea were planted by such large corporations as the very same General Electric and AT&T, with their state-of-the-art private research facilities. The work of such geniuses as Alexander Graham Bell and Thomas Edison demonstrated the utility, even the romance, of industrial research. Indeed, by the mid-1920s over 1,000 private companies maintained their own laboratories; American industry was spending twice as much money on basic research as was the federal government. (This preference for private over public expenditure was not necessarily shared by every industrial researcher. Willis R. Whitney, director of the General Electric Laboratory, told Hale that "while it is a good plan to ... encourage people of wealth to support the sciences ... it would be better if ... democracy could be brought to pay its own expenses in this line."[67] In other words, tax the chumps out in the hinterlands to pay for basic research rather than let philanthropists and industry pay the bill voluntarily. Mr. Whitney would get his wish, in due time.)

Why, wondered the ubiquitous George Ellery Hale, couldn't private businesses pool their resources to create a fund to sponsor researchers in pure science? In 1925, Hale convinced the National Academy of Sciences to create just such a fund in the National Research Endowment, which was soon renamed the National Research Fund.

The plan was for the NRF to be run by a board of trustees whose membership was blindingly bright: National Academy of Sciences eminentoes, statesmen, and sagacious men of business. Their number included former Secretary of State Elihu Root, titan Andrew Mellon, General Electric chairman Owen D. Young, Sears Roebuck chairman Julius Rosenwald, 1916 Republican Party presidential candidate and future Supreme Court Chief Justice Charles Evans Hughes, and defeated 1924 Democratic presidential candidate and Wall Street lawyer John W. Davis. The Chairman was Secretary of Commerce Herbert Hoover.

Hoover was neither the laissez-faire libertarian whose preferred course of action for the government was to do nothing, nor was he a proto-New Deal liberal. As the historian Ellis W. Hawley described Hoover's vision in an influential 1974 essay in the *Journal of American History*, he was an "associationist." That is, Hoover, an engineer who as head of the Food Administration during World War I had helped keep starvation at bay among the

Allies, saw a loosely knit network of "cooperative institutions, particularly trade associations, professional societies, and similar organizations among farmers and laborers" as the backbone of a healthy, community-oriented society that did not rely upon big or interventionist government for direction and succor.[68] In this way a system of social welfare might develop that was not reliant upon exorbitant taxes and productive of an intrusive national bureaucracy.

His pre-presidential platform for pushing this associative state was his position as Secretary of Commerce under President Harding. Though this might sound like something of a bureaucratic backwater, being "one of the smallest and newest of the federal departments," it was exactly where Hoover wanted to be; his intention was to make the Commerce Department the fulcrum "guiding the associational activities that were transforming American society."[69]

Hoover campaigned for the fund in a series of speeches arguing for "enlarged activities in support of pure science research."

The United States had scanted pure science, claimed Hoover. The nation had relied on three sources — the universities, benevolent men of wealth, and the rest of the world — to do the heavy scientific lifting for us. We underspent on pure research, and one result was a misallocation of human resources: "four or five thousand men in the United States" who were capable of performing research in the sciences were wasting their talents "applying themselves to the education of the youth." Thirty years hence, Americans would be lectured that they were falling behind the Soviet Union and imperiling the nation by not devoting *enough* resources to "the education of the youth," but then Hoover's themes and examples would reappear throughout the rest of the American Century, as partisans of government-funded science could argue these things any which way.

Genius cannot thrive in poverty, Hoover said. It needs expensive laboratory equipment. "The day of the genius in the garret has passed, if it ever existed," he said, not anticipating Silicon Valley.

The fault also lay in the frivolous preferences of Americans. "We spend more on cosmetics than we do upon safeguarding this mainspring of our future progress," asserted Hoover, presaging the "we spend more on potato chips than calculus" lectures of the 1950s and the 1960s.

Consistent with the advanced thinking of his times, Hoover looked in "three directions" for expanded support of basic research: "first, from the government both national and state"; second, "from industry"; and third, from "an enlargement of private benevolence." Cooperation of state, industry, and the wealthy was in the air: it was, in this rather benign form, a cousin of the fascist ideology that was spreading in Europe. This is not to say that Hoover was a fascist, only that his vision of a close connection between the central government and large industry was also a feature of fascist economics.

Yet he did believe in private industry, and he strained to emphasize that the National Research Fund was "no appeal for charity or benevolence" but instead "an appeal to self-interest." Pure science, in Hoover's words, was "the search for new fundamental natural law and substance," while applied science is the "application of these discoveries to practical use."[70] The applied research on which industry depended in turn depended upon basic research, and in subsidizing basic research, these industries were stocking their own larders.

The National Research Fund was supposed to be capitalized at $20 million. All monies invested would be spent within the year of the investment; this was not "philanthropy but a current expense, a payment for new knowledge and, as such, a way of obtaining a corporate benefit," as Lance E. Davis and Daniel J. Kevles wrote in their history of the fund in *Minerva*.[71]

As Davis and Kevles note, AT&T came on board with a pledge of $200,000 per annum for a decade; the National Electric Light Association followed up with a $3 million pledge, and Sears Roebuck's Julius Rosenwald pledged a cool million out of his own pocket. And then it stalled. The $20 million seemed like a distant and receding goal, especially since not a penny could be parceled out until the entire sum had been raised. So the goal was halved, but even $10 million was a tough struggle.

Then Herbert Hoover went and got himself elected president in 1928 — a promotion, it would seem, but the driving force behind the fund now had bigger fish to fry. U.S. Steel and the American Iron and Steel Institute came aboard, and despite the crash of 1929 George Eastman and the Rockefeller Foundation made large enough pledges to boost the total over the $10 million mark. Or so they thought. For when it came time to put their money where their mouth was, the members of the National Electric Light Association, a trade group, backed off. Making a $3 million pledge somehow seemed easier in 1927 than it did in 1930. The National Research Fund died.

Secretary of Commerce Hoover attributed the failure of businessmen to contribute to the NRF to the sad fact that "they have not grasped the essential differences between the applied science investigations upon which they are themselves engaged and the pure science which must be the foundation of their own future inventions."[72] In other words, blame that all-purpose villain who is always and ever to blame whenever someone fails to do what we believe he should do: Ignorance.

Eastman, Rosenwald, and the Rockefeller Foundation — private donors — had pledged 30 percent of the total. Not a single pledge came from the automobile, chemical, or railway industries. This was not because the directors of these companies were too dense to understand the importance of basic research. Indeed, chemical firms, report Davis and Kevles, "in 1927 employed about 3,400 persons in research — a figure second only to firms in electrical communication."[73] Rather, the stumbling block appeared to be the quite rational calculation that free riders — competitor firms that did not contribute to the fund — would nevertheless profit from its work, as the knowledge produced by fund researchers would be published and therefore available even to companies that had not put one red cent into the kitty.

It was, say Davis and Kevles, the awareness of these "externalities — the benefits which will accrue" to a firm's competitors — that doomed the NRF. "The campaign for the National Research Fund was an attempt to finance academic science in which those who paid the costs could not avoid having much of the resulting benefits flow to others," they write. As long as the funded scientists had the freedom to publish their results, as the NRF had guaranteed, there was simply no way that the funders could keep from aiding free riders. (Monopolies or companies with a dominant position worry less about such competition, and in fact, AT&T and U.S. Steel did contribute to the fund without having to worry overly much about aiding weak or nonexistent competitors.)

Unfortunately, the lesson drawn from the failure of the National Research Fund was not that a Hooverian "associative state" contained its own internal contradictions or that tax or copyright laws should be altered to encourage industrial research in basic science — which was, you will recall, flourishing as of the late 1920s anyway. No, the lesson, to quote Davis and Kevles, was that "Industrial corporations wishing to benefit from research could regard governmental support of academic research as a useful way of obtaining these benefits without the risks. Since the federal government would tax all corporations, there would be no externality of benefit to any firm in a given industry. Indeed, both businessmen and academic scientists urged federal support for the physical, or industrially relevant, sciences before the Second World War. In the battle over the establishment of the National Science Foundation after the Second World War, representatives of the National Association of Manufacturers strongly endorsed the creation of an agency organized in accord with the administrative principles and patent policies elaborated by Vannever Bush in *Science — the Endless Frontier*."[74]

Government was the tool by which the free rider problem could be solved. Moreover, by funding research out of the general treasury, the cost of such research could be passed along to shopowners, brewers, factory workers, insurance agents, and baseball players — the rest of the population. They would not benefit from this research, of course, at least not directly, but nor would the additional tax burden be so onerous as to rouse complaints. Far better it is to spread the burden around than to bear a disproportionate share of it — or so it might appear to a firm weighing the benefits of investing in the NRF.

"In the 1920s, astronomers expected to benefit from the National Research Fund," since its fathers included George Ellery Hale, and few were better than Hale at prying open pocketbooks."[75] But the NRF failed. The NSF would not fail — nor would it look to private industry for support.

President Truman signed the bill creating the National Science Foundation on May 10, 1950, while traveling by train in Idaho. The long quarrels over patent policy, and whether or not the social sciences should be included, ended in 1950, largely because America was embarking upon what would be a 40-year Cold War against the Soviet Union. Big government was back in the saddle. As J. Merton English writes in *A Patron for Pure Science: The National Science Foundation's Formative Years, 1945–1957* (1993), "The belief that the agency did have a supremely practical policy — the supplying of scientific and technical personnel and weaponry for national defense — probably did more than anything else to muster the votes needed for passage of the act of 1950."[76] James B. Conant, a Bush ally and president of Harvard, was the first chairman of the Board of NSF, and he was as political a scientist as they came.

However, in the five years that petty politics delayed the NSF, science did not stand still, human knowledge did not deteriorate, and the brains of physicists and astronomers and biologists did not atrophy. We would do well to remember this.

The National Science Board, appointed by President Truman and confirmed by the Senate, numbered 24. It first met December 12, 1950. One board member, California Institute of Technology President Lee A. DuBridge, called the foundation "the beginning of a New Deal for American science."[77] Its first batch of 28 grants were let in 1952 — a modest beginning. But not to worry. Its early years, said one observer in 1957, were marked by "lusty growth," a description to raise the envy of any bureaucrat.[78]

Even the house historian of the NSF, Milton Lomask, admitted that "The United States Constitution as ratified into life in 1789 did not in so many words permit Congress to finance scientific activity." It found a way, of course, entering through the general welfare clause. Vannevar Bush, adjudged Lomask, was "one of the major architects of the partnership now in effect between the American Government and the American scientific community."[79] Is this a partnership that the Founders envisioned, or that men and women of science should welcome?

To Vannevar Bush's disappointment, the NSF as finally created in 1950 was not the super-agency he had proposed, the nerve center of governmental scientific research. In fact, Bush "refused to allow researchers from the Carnegie Institution ... to accept foundation grants." He came to see it as a trough of pork, warning in 1953: "Concealed or disguised subsidy invites lobbying, puts a premium on scientific salesmanship – [and] might eventually prostitute the universities."[80] Yet by the late 1940s, as Daniel Kevles has written of Bush's old school, "85 percent of the MIT research budget came from the military and the AEC [Atomic Energy Commission]."[81] Pork was okay if it was sliced university-style.

So the NSF was up and running, fed by the Cold War. An assistant secretary of defense in the Eisenhower administration, C.C. Furnas, said in 1956, "The tremendous military research and development program will continue as far as I can see, for a long time in the future. It furnishes the background and a large part of the stimulus for this rapidly rising truly scientific age."[82]

Nobel Prize winning physicist I.I. Rabi of Columbia University, looking back from the vantage point of the 1980s on the Cold War matchup of the state and the university, said, "The whole picture changed as a result of government money.... It stopped being a collegial affair. I thought it was a wonderful thing ... but the costs of it, which were great, I didn't realize until later."[83]

The ideas of such men as Frank Jewett were buried in the rubble of the 1950s. As historian Stuart W. Leslie writes, "For better and for worse, the Cold War redefined American science. In the decade following the Second World War, the Department of Defense (DOD) became the single biggest patron of American science, predominantly in the physical sciences and engineering but important in many of the natural and social sciences as well."[84]

A new war, though much smaller than the Second World War, helped to cement the new scientific order. "The Korean War completed the mobilization of American science and made the university, for the first time, a full partner in the military-industrial complex," according to Leslie, as "Military money flooded into industrial and academic laboratories." By the 1960s, such elite schools as MIT, Stanford, and Cal Tech had become virtual adjuncts of the federal government, so dependent were they upon federal research contracts. Of MIT, physicist Alvin Weinberg joked in 1962, one could hardly "tell whether the Massachusetts Institute of Technology is a university with many government research laboratories appended to it or a cluster of government research laboratories with a very good educational institution attached to it."[85] (For more on the close postwar linkage between elite science-engineering schools and the government, see Leslie's *The Cold War and American Science* (1993).)

In time, scholars would cast a critical eye on the sea change. In *A Fragile Power: Scientists and the State* (1989), Chandra Mukerji noted that in all the quibbling over which are the best strategies to use to pry more money out of government for science, "no one asks what scientists give up to get government funds."[86]

The tradeoff, writes Mukerji in her study of the culture of ocean scientists, may not be worth it. "Science gains much of its financing and most of its social power because of its usefulness to government, but science cannot easily prosper as an intellectual endeavor simply by serving the powerful on issues of their choosing." In accepting massive amounts of government funding, one accepts a kind of de facto government direction. "Agencies such as the NSF help to define for scientists and politicians alike what constitutes quality research," she writes. Scientists may kid themselves that their pure research is being conducted on the lofty plane of unimpeded intellect and discovery, but they have been elevated to that plane not by the invisible hand of the marketplace but the intrusive hand of the state. The *funded* are at the mercy of the *funders*. And when that funder is the state, the funded are, for all intents and purposes, servants of the state. The system, Mukerji argues, "serves the state more than science."[87]

This was a drastic alteration of the American system. As Roger L. Geiger writes in *Knowledge & Money: Research Universities and the Paradox of the Marketplace* (2004), "Historically ... European universities were traditionally accorded comfortable support by the state to cultivate and profess learning. Private patrons similarly supported American universities before World War II, if more irregularly."[88] Postwar, everything changed. The federal government supplanted private parties as patrons of academic research. This research was disproportionately directed toward state concerns — defense, especially — but the universities weren't complaining as long as the money flowed.

As Geiger notes, "by the time NSF was created in 1950, federal commitments to academic science were already dominated by the requirements of national defense and, to a lesser extent, public health."[89] *Sputnik* would later add another justification — winning the space race, and the education "war" with the Soviets.

Chandra Mukerji speaks of government-funded scientists as "a reserve labor force," not completely unlike welfare recipients. "[L]ike the unemployed on welfare," she writes provocatively, scientists on research grants are "kept off the streets and in good health because of the interests and investments by elites." They are kept alive, professionally speaking, in case the state needs them in the event of war. For the lesson of the Second World War and Cold War was that science, more than mere manpower, is the key to military victory.

War's end presented all sorts of messy conversion problems. As Mukerji writes, "Scientists would have to be demobilized, but they did not have to abandon their pursuit of strategically valuable skills acquired during the war. They could continue their research on grants or contracts, while working in the university or private research institution."[90] All that was needed was the infusion of federal cash. It was coming — though it hardly met anyone's idea of a purity standard, since nine of every ten dollars the U.S. was spending on R&D by the late 1940s was supplied by the Department of Defense and the Atomic Energy Commission. And as Kevles says, "the defense R&D budget followed the overall defense budget into the stratosphere" in the Korean War era — a development that the president who was elected to get us out of Korea, Dwight Eisenhower, would later wrestle with. Even such a supposedly pure science as particle physics was a "protégé of national security," as the accelerator labs were born of the Cold War.[91]

These complications were not unanticipated at the birth of the NSF. The clearest exposition of the problem came from.... Frank Jewett, of course. "We know and are dazzled by the accomplishments of applied science mobilized for a single purpose — war," said Jewett in the interval between World War II and Korea. Yet Jewett knew that the conditions which produced these dazzling accomplishments were not only unduplicable but they had also come at a serious cost to American science, for "we have lost irrevocably the better part of a generation of creative research men and the better part of a generation of creative additions to our stock pile of fundamental knowledge." Men and women who might have contributed to the advance of science were detoured into government work, into the application of knowledge for the building of weaponry. The universities, drained of young males, could not do their jobs. And of course many who might have contributed to the growth of human knowledge were killed in the bloodbath. He worries that the habits of regimentation and obedience inculcated in wartime may mean that "science will continue to be the servant of political government." For "No matter what plausible arguments are advanced, fundamental science cannot flourish in peacetime under the regimentation of a wartime setup or under bureaucratic control."

Jewett takes on, directly, the legislative spawn of Vannevar Bush. "All the proposals for federal support of science now being considered involve some measure of direct or indirect political control of science," especially S. 1850, the Kilgore bill, which involves "strong controls which are far reaching, uncertain, and dangerous." It is all very well for scientists to welcome the inflow of federal dollars, he says, but the "things that are not so obvious are the restraints on free scientific research which the dispensing of such money would inevitably place in the hands of political government and in an unknown, essentially uncontrolled bureaucracy."

Jewett asked the tough questions that no one else was asking. Sure, the infusion of massive amounts of federal dollars would stimulate "progress" in certain areas of research, but he questioned "the tacit assumption that it is clearly in the interest of society greatly to stimulate the progress of science in all sectors far beyond that which would occur if things were left to the normal balances of interest in a complex society." Might not a laissez-faire approach in which research unfolded more naturally, responding to the direction of free men and women in a society based more on voluntary cooperation than coercion, produce a science that conformed better to human nature? Was it possible that the gadgets and weapons manufactured by state directed science might make us less free, or subordinate traditional American liberties to the demands of big government? Were constant panics and wars really the best goads to research?

Jewett hastened to say that he was not advocating "even a partial holiday in scientific research" — he just thought that research should not be artificially stimulated by the all-powerful central government. Do not forget freedom, Jewett pled. The "fruits" of "fundamental science," he said, "are those of the free mind and no one is wise enough to know what another man's brain cells may produce if afforded opportunity to function freely." In contrast to wartime bureaucracy, said Jewett, the peacetime industrial laboratory offers minimal direction and maximum freedom to researchers.

Jewett was not optimistic that American science could rebound from the war. Government profligacy and obedience to central authorities had been the dominant themes since 1941. "War," he said, "destroys those standards of thrift, frugality, and efficiency of effort which are necessary in a productive civil economy. We have just been through a long experience where waste, squandered money, and lack of efficiency have come to be taken as a matter of course. In the field of science this had bred in many young men (and many older ones too) a feeling that nothing worth while can be accomplished without elaborate facilities, a wealth of assistance, and a lack of desire to plan economically which will continue to plague us in our endeavor to get back on an even keel. We will have demands for more and more elaborate equipment, for more assistance on the plea that work can't be done otherwise, and, in the field of education, more pleas for scholarship aid from those who before the war would not have expected it."[92]

This is exactly what we got.

By 1946, Warren Weaver, program director of the Rockefeller Foundation, despaired that the "magnitude of the federal support, as opposed to anything that could reasonably come from the Rockefeller Foundation," was marginalizing private and nonprofit patronage of scientists.[93] The old arrangement had been routed.

One of the most cogent and searching critiques of the new arrangement came from an unexpected source: President Dwight D. Eisenhower.

In his Farewell Address, delivered January 17, 1961, on the eve of a new administration full of pep and verve and style and a seemingly boundless faith in the abilities of the military and science to conquer anything, President Eisenhower sounded an ominous warning that might have come from American voices of the past:

> Today, the solitary inventor, tinkering in his shop, has been overshadowed by task forces of scientists in laboratories and testing fields. In the same fashion, the free university, historically the fountainhead of free ideas and scientific discovery, has experienced a revolution in the conduct of research. Partly because of the huge costs involved, a government contract becomes virtually a substitute for intellectual curiosity. For every old blackboard there are now hundreds of electric computers.
>
> The prospect of domination of the nation's scholars by federal employment, project allocations, and the power of money is ever present – and is gravely to be regarded.
>
> Yet, in holding scientific research and discovery in respect, as we should, we must also be alert to the equal and opposite danger that public policy could itself become the captive of the scientific-technological elite.[94]

This was an extraordinary warning. Coupled with Eisenhower's strictures upon the "military–industrial complex," it was as if the old General, who had spent a lifetime in the employment of the national government, had woken up to the incredible spread of the federal government into fields that had once seemed to be wholly or mostly private. And he wished to use his last speech to the American people to issue a profound caution.

It almost seemed too late: the horse was out of the barn. "By the mid-1960s, the federal government had come to fund 80 percent of all research and development performed in the United States," writes Walter McDougall.[95] The universities were becoming adjuncts of Washington: almost 90 percent of Cal Tech's research budget came from the federal government.

Scientists were not blameless in this. As scholar H.M. Sapolsky charged, in "their enthusiasm for research subsidies and their willingness to accept whatever rationales appeared to be effective in gaining such subsidies," scientists "have produced precisely the impact upon science they sought to avoid   its permanent mobilization."[96] He who takes the king's shilling becomes the king's man. But if there are enough shillings, men (and women) will convince themselves that somehow *they* are not part of the problem, they are uncorrupted by the kingly eleemosynary, and that in fact by taking the king's shilling they may even be furthering the anti-monarchist cause. When fat grants are on the line, people are capable of convincing themselves of many things.

Consider Jerome B. Wiesner, who served as Special Assistant for Science and Technology to JFK and LBJ, and who was not buying the Eisenhower line about the dangers to science posed by government dependence and the military–industrial complex.

Wiesner, awed by what he called "the incomprehensible power of the office" of the president, boasted that "Government support for research in the universities is one of the prime reasons for their greatness in science and engineering."[97] There was, it seems, no limit to the scope of state patronage of science. (In 1962, in the words of Simon Ramo, "the electronics industry has become greatly dependent now on government support."[98] What wasn't?)

Spurred by wars, hot and cold, by 1963 the federal government was responsible for 12 of the 16 billion dollars spent on scientific research and development in the United States. Wiesner views that development without regret. During the Second World War, he says, "a pattern of military–industry–university interdependence grew into being," and though a brief demobilization followed, "In response to the Korean War, the amount of federally supported research and development grew again; it since has sustained an almost continuous growth."[99] By the New Frontier of JFK, the sky was the limit — and as we shall see in the next chapter, the sky and the scary things flying around up there would spur the federal government to even greater expenditures.

As for the NSF, it has now spent 60 years avoiding the question that Charles Kidd, writing in 1959, posed: "how federal participation is to be achieved without federal domination."[100]

Federal sponsorship can subtly, or not so subtly, nudge the direction of scientific research down some channels and away from others. Joseph P. Martino, in his study *Science Funding: Politics and Porkbarrel* (1992), writes, "In short, the dominant influence of the NSF in certain disciplines means that a handful of individual project officers in the NSF can shape the content of entire disciplines."[101] These officers are not, as Martino emphasizes, necessarily ogres or mini-tyrants or power-mad megalomaniacs who want to alter the path of scientific research. They are human beings doing the best they can. But they have the power to channel scientific research down paths of the state's choosing.

This is an unavoidable peril of any centralized funding system — indeed of any government funding that comes from Washington. Francis Crick, Nobel laureate and codiscoverer of the structure of DNA, said in 1987: "Money for research has to come from somewhere, be it robber barons or the taxpayer. The best way to distribute it is not through some monolithic system, however much care is taken in choosing the right recipients. This is always fallible and can waste scientists' time interminably, sitting on tedious committees. Far better to have many sources of money, with a series of minidictators to distribute it."[102]

This runs contrary to the mindset of the bureaucratic funder of scientists, but then what does a mere Nobel laureate know? Scientists do have troublesome independent streaks, in the eyes of those who would control their pursestrings.

Once firmly established, the NSF no longer had to deal with any Frank Jewetts questioning its very existence. "The market mechanism by itself is likely to lead to an underinvestment in research and development from society's point of view," authoritatively stated the 1978 National Science Foundation annual report.[103] Jewett would have denied the premise. He would have marshalled almost two centuries' worth of evidence in his favor. But by 1978, Vannevar Bush and Harley Kilgore, different as those two men might have been, had carried the day.

When President Obama addressed the annual meeting of the National Academy of Sciences in April 2009, he upped the ante, as presidents are wont to do when currying favor with an audience, saying of his science research-related spending plans that "We will not just meet but we will exceed the level achieved at the height of the space race."[104] It went without saying that this money was to be spent not by private research labs, the General Electrics of those bygone days of Frank Jewett and L.A. Hawkins, or by the philanthropically endowed observatories or laboratories that once stood as monuments to the potential of privately funded research. Instead, President Obama promised to double the budgets of the various science agencies within the federal leviathan, which span the bureaucracy from the Department of Defense to the Department of Energy and whose spearhead, which the President mentioned with the reverence due an old and honorable warhorse in the collective pursuit of knowledge, is the National Science Foundation.

The Democratic budget boosted the NSF budget from $6 billion to $9 billion over the course of the first twenty months of the new Obama administration, making it, as *Science* magazine's website put it, "the biggest winner, on a percentage basis, of any research agency" under the newly ascendant Democrats. The NSF's congressional affairs spokesman termed it "an incredibly positive message [from Congress] about the value of basic science in contributing to economic recovery. It's unbelievable."[105]

Our forebears would have found "unbelievable" the concept of scientific research directed by a central state. Even our immediate predecessors — with obvious exceptions such as Senator Harley Kilgore, who must have been doing flips of joy in his grave — would have found "unbelievable" the notion that scientific research should be integral to an economic stimulus package, the equivalent, say, of New Deal pump-priming and make-work initiatives. And when one looks at certain of the projects that the NSF boost was going to fund — the Ocean Observatories Initiative, the National Ecological Observatory Network, the Alaska Regional Research Vessel, and most of all the Advanced Technology Solar Telescope — one cannot help but reflect that these were projects that in an earlier stage of the republic would have been well-funded by the private sector. Especially the solar telescope. Are there no longer rich people in America eager to attach their names to such projects? Ask Bill Gates.

Joseph P. Martino has written, "We have become so used to federal support of research, with industrial support as a second option, that it is sometimes hard to remember that there was even any other way. Yet before World War II most American science was supported by private funds, and prior to World War I virtually all American science was so supported."[106]

No one has forgotten that lesson more thoroughly than the bureaucrats whose redoubt was established just after the Second World War, when there seemed nothing that government, having won the war, couldn't do.

Well into its second half century of life, the National Science Foundation shows no signs of shrinkage. Its champions no longer feel obligated to pay lip service to private science, to unsubsidized researchers. As the NSF slogan these days goes: the National Science Foundation is "Where Discoveries Begin."

Notes

1. Paul Forman, "Behind Quantum Electronics: National Security as Basis for Physical Research in the United States, 1940–1960," *Historical Studies in the Physical and Biological Sciences*, Vol. 18 (1987), p. 152.
2. Milton Lomask, *A Minor Miracle: An Informal History of the National Science Foundation* (Washington, DC: National Science Foundation, 1975), p. 35.
3. Leslie, *The Cold War and American Science: The Military-Industrial-Academic Complex at MIT and Stanford*, pp. 5–6.
4. Simon Rottenberg, "The Economy of Science: The Proper Role of Government in the Growth of Science," *Minerva*, Vol. 19, No. 1 (March 1981), p. 56.
5. Merton England, *A Patron for Pure Science: The National Science Foundation's Formative Years, 1945–1957* (Washington, DC: National Science Foundation, 1983), p. 7.
6. Forman, "Behind Quantum Electronics: National Security as Basis for Physical Research in the United States, 1940–1960," p. 157.
7. A. Hunter Dupree, "The Great Instauration of 1940: The Organization of Scientific Research for the War," in *The Twentieth-Century Sciences: Studies in the Biography of Ideas*, edited by Gerald Holton (New York: W.W. Norton, 1972), p. 457.
8. Ibid., p. 462.
9. Daniel J. Kevles, "The National Science Foundation and the Debate over Postwar Research Policy, 1942–1945: A Political Interpretation of *Science — The Endless Frontier*," *Isis*, Vol. 68, No. 1 (March 1977), p. 12.
10. Vannevar Bush, *Science — The Endless Frontier* (Washington, DC: National Science Foundation, 1960/1945), p. 1.
11. G. Paschal Zachary, *Endless Frontier: Vannevar Bush, Engineer of the American Century* (New York: Free Press, 1997), pp. 218–19.
12. Alan T. Waterman, "Introduction," *Science — The Endless Frontier*, p. vii.
13. Bush, *Science — The Endless Frontier*, pp. 3–4.
14. Waterman, "Introduction," *Science — The Endless Frontier*, p. vii.
15. Bush, *Science — The Endless Frontier*, p. 2.
16. Ibid., pp. 5, 8.
17. Ibid., 31.
18. Ibid., p. 8.
19. Ibid., p. 12.
20. Ibid., p. 17.

21. Waterman, "Introduction," *Science — The Endless Frontier*, p. vii.
22. Bush, *Science — The Endless Frontier*, p. 32.
23. Ibid., p. 20.
24. Zachary, *Endless Frontier: Vannevar Bush, Engineer of the American Century*, pp. 221–22, 224.
25. Bush, *Science — The Endless Frontier*, p. 74.
26. Ibid., pp. 83–84, 77–78.
27. Quoted in Kevles, "The National Science Foundation and the Debate over Postwar Research Policy, 1942–1945: A Political Interpretation of *Science — The Endless Frontier*," p. 19.
28. Bush, *Science — The Endless Frontier*, pp. 91–92.
29. Ibid., p. 75.
30. Ibid., pp. 96, 93–94.
31. Zachary, *Endless Frontier: Vannevar Bush, Engineer of the American Century*, pp. 4, 295.
32. Bush, *Science — The Endless Frontier*, p. 115.
33. Ibid., p. 35.
34. Zachary, *Endless Frontier: Vannevar Bush, Engineer of the American Century*, pp. 232–33.
35. *The Politics of American Science: 1939 to the Present*, edited by James L. Penick Jr, Carroll W. Pursell Jr., Morgan B. Sherwood, and Donald C. Swain (Chicago: Rand McNally & Co., 1965), p. 35.
36. Quoted in Kevles, "The National Science Foundation and the Debate over Postwar Research Policy, 1942–1945: A Political Interpretation of *Science — The Endless Frontier*," p. 11.
37. *The Politics of American Science: 1939 to the Present*, pp. 42–44.
38. Waldemar Kaempffert, "Horizons of Science," *The American Mercury* (October 1943), p. 441–47.
39. American Association of Scientific Workers collection, http://library.temple.edu/collections.
40. L.A. Hawkins, "Regimentation of Science," *Electrical World* (June 26, 1943), pp. 45–47.
41. "A National Science Program," *The New Republic* (July 30, 1945), p. 116.
42. Bush, *Science — The Endless Frontier*, pp. 85–86.
43. Kevles, "The National Science Foundation and the Debate over Postwar Research Policy, 1942–1945: A Political Interpretation of *Science — The Endless Frontier*," p. 7.
44. *The Politics of American Science: 1939 to the Present*, pp. 54, 57.
45. Zachary, *Endless Frontier: Vannevar Bush, Engineer of the American Century*, pp. 252–53.
46. Kevles, "The National Science Foundation and the Debate over Postwar Research Policy, 1942–1945: A Political Interpretation of *Science — The Endless Frontier*," p. 23.
47. Zachary, *Endless Frontier: Vannevar Bush, Engineer of the American Century*, pp. 252, 328.
48. *The Politics of American Science: 1939 to the Present*, pp. 87–88.
49. Quoted in ibid., p. 82.
50. James D. Savage, *Funding Science in America: Congress, Universities, and the Politics of the Academic Pork Barrel* (Cambridge: Cambridge University Press, 1999), p. 36.
51. Oliver E. Buckley, *Frank Baldwin Jewett 1879–1949* (Washington, DC: National Academy of Sciences, 1952), p. 244.

52. *Science in America: Historical Selections*, edited by John C. Burnham (New York: Holt, Rinehart and Winston, 1971), p. 398.
53. Lomask, *A Minor Miracle: An Informal History of the National Science Foundation*, p. 46.
54. "Testimony of Dr. Frank B. Jewett," Hearings on Science Legislation S. 1297 and Related Bills, U.S. Senate Committee on Military Affairs, Subcommittee on War Mobilization, Washington, DC, pp. 427–47, *passim*.
55. Quoted in *The Politics of American Science: 1939 to the Present*, pp. 85–86.
56. Quoted in Lomask, *A Minor Miracle: An Informal History of the National Science Foundation*, p. 48.
57. England, *A Patron for Pure Science: The National Science Foundation's Formative Years, 1945–1957*, pp. 29, 25, 10.
58. Lomask, *A Minor Miracle: An Informal History of the National Science Foundation*, p. 56.
59. England, *A Patron for Pure Science: The National Science Foundation's Formative Years, 1945–1957*, pp. 78, 59, 48.
60. Quoted in ibid., p. 96.
61. Quoted in ibid., pp. 36, 45–46.
62. Quoted in ibid., p. 381.
63. John R. Baker, *Science and the Planned State* (London: George Allen & Unwin Ltd., 1945), pp. 9, 15, 48.
64. Quoted in ibid., pp. 11, 16.
65. Ibid., p. 41.
66. Martino, *Science Funding: Politics and Porkbarrel*, p. 229.
67. Lance E. Davis and Daniel J. Kevles, "The National Research Fund: A Case Study in the Industrial Support of Academic Science," *Minerva*, Vol. 12 (April 1974), p. 220.
68. Ellis W. Hawley, "Herbert Hoover, the Commerce Secretariat, and the Vision of an 'Associative State,' 1921–1928," *Journal of American History*, Vol. 61, No. 1 (June 1974), p. 117.
69. Ibid., pp. 120, 119.
70. Herbert Hoover, "The Nation and Science," *Science*, Vol. 65, No. 1672 (January 14, 1927), pp. 26–29.
71. Davis and Kevles, "The National Research Fund: A Case Study in the Industrial Support of Academic Science," pp. 210–11.
72. Hoover, "The Nation and Science," p. 28.
73. Davis and Kevles, "The National Research Fund: A Case Study in the Industrial Support of Academic Science," p. 214.
74. Ibid., pp. 217, 219.
75. Lankford and Slavings, "The Industrialization of American Astronomy, 1880–1940," p. 40.
76. England, *A Patron for Pure Science: The National Science Foundation's Formative Years, 1945–1957*, p. 7.
77. Lomask, *A Minor Miracle: An Informal History of the National Science Foundation*, p. 31.
78. England, *A Patron for Pure Science: The National Science Foundation's Formative Years, 1945–1957*, p. 347.
79. Lomask, *A Minor Miracle: An Informal History of the National Science Foundation*, pp. 33, ix.

80. Zachary, *Endless Frontier: Vannevar Bush, Engineer of the American Century*, pp. 369–70.

81. Forman, "Behind Quantum Electronics: National Security as Basis for Physical Research in the United States, 1940–1960," p. 157.

82. Quoted in ibid., p. 149.

83. Quoted in ibid., p. 181.

84. Leslie, *The Cold War and American Science: The Military-Industrial-Academic Complex at MIT and Stanford*, p. 1.

85. Ibid., pp. 8, 14.

86. Chandra Mukerji, *A Fragile Power: Scientists and the State* (Princeton: Princeton University Press, 1989), p. x.

87. Ibid., pp. 4, 6, 104.

88. Roger L. Geiger, *Knowledge & Money: Research Universities and the Paradox of the Marketplace* (Stanford: Stanford University Press, 2004), p. 2.

89. Ibid., p. 136.

90. Mukerji, *A Fragile Power: Scientists and the State*, pp. 7–8.

91. Dan Kevles, "Cold War and Hot Physics: Science, Security, and the American State, 1945–56," *Historical Studies in the Physical and Biological Sciences*, Vol. 20, No. 2 (1990), pp. 250, 263.

92. Frank B. Jewett, "The Future of Scientific Research in the Postwar World," in *Science in America: Historical Selections*, pp. 399–413, *passim*.

93. Quoted in Forman, "Behind Quantum Electronics: National Security as Basis for Physical Research in the United States, 1940–1960," p. 186. Eventually, some scholars dared to peer into the science pork barrel. James D. Savage, who had studied academic earmarking for the Congressional Research Service in the early 1990s, wrote in 1999, "Sixty percent of academic research is funded by the federal government," and while peer or merit review, the preferred methods for allocating such dollars, have their own problems — namely, favoritism and a tendency to support those with conventional views over scientific rebels — academic earmarking by members of Congress brings to science all the elegance and subtlety of the political pork barrel. Yet peer review does indeed favor elite institutions and can be just as ridden with good old boy networks as a Chicago road-building appropriation. As Savage explains the critique, "Peer review, it is claimed, is an old boys' network that helps the rich get richer. It rewards the same old schools, the same old scientists, and the same old science, and thereby denies funding for new and innovative research efforts for scientists other than the privileged few." Savage, *Funding Science in America: Congress, Universities, and the Politics of the Academic Pork Barrel*, pp. 5–7.

94. Dwight D. Eisenhower, "Farewell Radio and Television Address to the American People," January 17, 1961, The American Presidency Project, www.presidency.ucsb.edu/ws.

95. McDougall, "Technocracy and Statecraft in the Space Age — Toward the History of a Saltation," p. 1032.

96. Quoted in Forman, "Behind Quantum Electronics: National Security as Basis for Physical Research in the United States, 1940–1960," p. 226.

97. Jerome B. Wiesner, *Where Science and Politics Meet* (New York: McGraw-Hill, 1965), pp. 13, 17.

 98. Quoted in Forman, "Behind Quantum Electronics: National Security as Basis for Physical Research in the United States, 1940–1960," p. 166.
 99. Wiesner, *Where Science and Politics Meet*, p. 43.
100. Quoted in Martino, *Science Funding: Politics and Porkbarrel*, p. 85.
101. Ibid., p. 88. In the early 1980s, a lawsuit by California Rural Legal Assistance (CRLA) shed an interesting light on the question of how government-subsidized research advantages some and disadvantages others. The CRLA, acting on behalf of nineteen farm workers, filed suit against the University of California, charging that the UC had subsidized agricultural research with a "basic policy goal" of developing "machines and other related technology in order to reduce to the greatest extent possible, the use of labor as a means of agricultural production." In other words, the activists alleged that the state was subsidizing mechanization and the consequent elimination of manual labor farm jobs, especially for those who aided in the planting and harvesting of tomatoes, grapes, oranges, peaches, and lettuce. Not being of a particularly libertarian bent, the CRLA did not merely demand that the state refrain from subsidizing mechanized agriculture; instead, it requested that the University of California set up a retraining fund for such workers, to be paid out of a fund which would be capitalized by licensing and royalty payments from the inventions of its researchers. Whether or not this mechanization was a good or bad thing is not the point here. Philip L. Martin and Alan L. Olmstead, professors of agriculture at the University of California-Davis, argued in *Science* that "Mechanization reduces the arduous nature of harvest work and permits remaining farm workers to operate equipment and sort commodities for longer periods," while critics claim that it pushes farm workers out of productive outdoor employment and onto the unemployment or welfare rolls. Wherever the truth lies, it is unarguable that publicly funded — as distinguished, crucially, from privately funded — mechanization research is a putatively "neutral" state intervention in applied science whose consequences are anything but neutral. In fact, these interventions privilege certain actors and harm others. Even in a relatively uncontroversial field such as agricultural research, state-subsidized science has broad ramifications. Whether or not one thinks that the applied research is a boon to the agricultural economy, it is anything but "neutral." See Philip L. Martin and Alan L. Olmstead, 'The Agricultural Mechanization Controversy," *Science*, Vol. 227, No. 1687 (February 8, 1985), pp. 601–606.
102. Quoted in Martino, *Science Funding: Politics and Porkbarrel*, p. 372.
103. Quoted in Rottenberg, "The Economy of Science: The Proper Role of Government in the Growth of Science," p. 57.
104. Randolph E. Schmid, "Obama Promises Major Investment in Science," Associated Press, April 27, 2009.
105. Jeffrey Mervis, "A $3 Billion Bonanza for NSF?" *Science Insider*, January 16, 2009, http://blogs.sciencemag.org/scienceinsider.
106. Martino, *Science Funding: Politics and Porkbarrel*, p. 301.

Chapter 4

Oh *Sputnik!* How the Educationists Prospered from a Russian Satellite

On October 4, 1957, the Soviet *Sputnik I*, a 184 pound artificial satellite, orbited the earth. This was not wholly unexpected; watchers of the skies and the Soviets knew that the USSR and the United States had hoped to send such projectiles into space to mark the International Geophysical Year (IGY). The Soviets got there first with the "Artificial Fellow Traveler Around the Earth," as *Sputnik* was clumsily translated.

American scientists who learned of *Sputnik* while sipping vodka and limes at a reception at the Soviet Embassy in Washington in honor of the IGY were happy for the achievement; their national pride was not particularly bruised. "We are all elated that it is up there," remarked one.[1]

Sputnik hardly appeared de novo, as if revealed by a Soviet magician. As Tom Wolfe writes in *The Right Stuff* (1980), the "idea of an artificial earth satellite was not novel to anyone who had been involved in the rocket program at Edwards" Air Force Base "So what was the big deal about *Sputnik 1*?" thought the test pilots. But as Wolfe writes, men like Chuck Yeager did not reckon on the power of "technological illiterates with influence.... It was hard to realize that *Sputnik 1*... would strike terror in the heart of the West."[2]

President Eisenhower, seemingly alone among Washington figures, was unpanicked by *Sputnik*. He played golf five times that week, taking a good deal of heat in the press for it. He congratulated Soviet scientists via press release. In fact, Ike played a round of 18 holes the very next afternoon at Gettysburg (PA) Country Club — just as he had the day before.[3]

Putting the achievement in perspective, Eisenhower, in his press conference of October 9, 1957, responded to a question from Mrs. May Craig of the

J.T. Bennett, *The Doomsday Lobby: Hype and Panic from Sputniks, Martians, and Marauding Meteors*, DOI 10.1007/978-1-4419-6685-8_4,
© Springer Science+Business Media, LLC 2010

Portland (ME) *Press Herald*: "the Russians, under a dictatorial society where they had some of the finest scientists in the world who have for many years been working on this, apparently from what they say they have put one small ball in the air."[4] Whoopty-do, in other words.

Ike's Secretary of Defense, Charles E. Wilson, said it was a "nice scientific trick."[5] His equally budget-conscious Treasury Secretary, George Humphrey, said, "The real danger of the *Sputnik* is that some too-eager people may demand hasty action regardless of cost in an attempt to surpass what the Russians have done."[6]

President Eisenhower battled manfully to stem the propaganda tide, but within very short order bureaucrats, military men, union leaders, and others seeking federal favors had realized that exaggerating Soviet capabilities and the pathetic state of America was the key to opening the treasury vault.

Newspapers tried to fan a panic with headlines about the Russian "moon" that might be keeping an eye on sleeping Americans. (It wasn't.) An exception was the conservative *U.S. News & World Report*, whose editor, David Lawrence, commended Eisenhower for his "courageous statesmanship" in refusing to panic. Scientists who favored an expanded government role in the knowledge industry played up the threat. Edward Teller, associate director of the Livermore Radiation Laboratory, said that the U.S. had "lost a battle more important and greater than Pearl Harbor." German emigre rocket scientist Werner von Braun engaged in a bit of Cold War hyperbole when he warned that Americans might be "surrounded by several planets flying the hammer and sickle flag."

Hawkish politicians had a field day castigating Eisenhower and calling for new infusions of tax dollars. Senator Henry "Scoop" Jackson (D-WA) proclaimed a "National Week of Shame and Danger."[7] However, Cal Tech president Lee DuBridge had enough of this nonsense about control of the moon equaling control of the earth: "Why transport a hydrogen warhead together with all men and equipment 240,000 miles to the moon, just to shoot it 240,000 miles back to earth, when the target is only 5,000 miles away in the first place?"[8] He received no answer to his question.

The hysteria was unquelled.

Democrats, frustrated by two consecutive defeats for the presidency, spoke from the Democratic Advisory Council, whose ranks included former President Harry Truman, saying that *Sputnik* "must be met by all-out efforts of our own."[9] These "all-out efforts" would cost money, lots of it, and erode federalism, but the skies were falling, or at least bleeding communist red.

Grasping at their default setting, Senators Hubert Humphrey (D-MN), Stuart Symington (D-MO), and Estes Kefauver (D-TN) proposed a

Department of Science and Technology. Sure, creating a Cabinet-level science bureaucracy would get that American satellite built in a jiffy! Ike, not a rockets-and-spacesuits enthusiast, held back.

The specter of Soviet parity, even domination, of space haunted Americans, or at least it haunted their political leaders, notably such hawkish Democrats as Scoop Jackson and Senator Lyndon Baines Johnson (TX), who never met a taxpayers' dollar that they did not think was better spent by Congress than by a mere taxpayer. Johnson, who was even more unpleasant to be around than usual in the days following the launch, warned, "Now the Communists have established a foothold in outer space. It is not very reassuring to be told that next year we will put a 'better' satellite into the air. Perhaps it will even have chrome trim and automatic windshield wipers."[10]

This swipe at American automobile technology (which led the world) and the consumers who enjoyed such extras and amenities instead of being satisfied with dreary Eastern European-style Pobeda automobiles presaged a whole new line of attack by Big Science advocates who were outraged that Americans were spending their money on hoop skirts and sock hops and hot rods instead of government-directed science projects.

The New Republic, flagship of the establishment liberals, joined the panic: *Sputnik* was "proof of the fact that the Soviet Union has gained a commanding lead in certain vital sectors of the race for world scientific and technological supremacy."[11] It was no such thing, any more than a gold medal by a Jamaican sprinter in the Olympics was "proof" that the Jamaicans had overtaken the Americans and Russians in national sports supremacy.

Among the first lessons of *Sputnik*: the U.S. wasn't spending nearly enough on science education, and only the federal government could ensure that it did. Newspapers were filled with stories of the incredible efficiency of the Soviet educational system and the sluggishness of the American one. Benjamin Fine of the *New York Times*, who was always happy to peddle the conventional wisdom in urgent prose, highlighted the Soviet–American gap. The implication was that we needed to be more like the Russians and cede control of education to the central state, or else the sky would be filled with *Sputniks* raining death on America. *Life* magazine pointed to the Soviet tenet that education "draws or forces all human knowledge into service of the state."[12] Was this superior to American laissez-faire?

The hero of *Sputnik*, though he was vilified at the time, was President Dwight D. Eisenhower, who understood it as a flashy sideshow. If indeed there was a "race" between the Soviet Union and the United States, the U.S. was far in the lead by every measure that counts.

Yet the panic had a momentum of its own, and as Cold War historian Robert A. Divine points out, the President "finally was forced to go against his deeply ingrained fiscal conservatism and approve defense expenditures he did not believe were really needed." These included expenditures on a hectic space race that President Eisenhower believed to be misguided.

Hyperbole was the order of the day, or the year, or the era. The Middle American who was working hard and trying to improve his lot was to blame for *Sputnik.* As New Hampshire Republican Senator Styles Bridges – a man whose reported ethical lapses could fill a book – said, "The time has clearly come to be less concerned with the depth of pile on the new broadloom rug or the height of the tail fin on the car and to be more prepared to shed blood, sweat and tears if this country and the Free World are to survive."

Unlike his successor, Eisenhower insisted that the U.S. was not in a "race with other nations" to see which could launch satellites first. Science was not a competitive sport to be played like touch football in Hyannis Port. The Soviets had put "one small ball in the air." Fine, congratulations.

Edward Teller, who never failed to seize an opportunity to call for more federal spending on matters of interest to him, said gravely, "We have suffered a very serious defeat in a field where at least some of the most important engagements are carried out: in the classroom." Another physicist, George R. Price of the University of Minnesota, chimed in that selfish Americans were going to have to be more concerned with "giving our children a good education than about keeping our property taxes low."[13]

The chorus of establishment wisdom sang a new song: Schools in Russia were superior to those in America. Young Russians were being trained as scientists and engineers, careers vital to success in the Cold War, while American schools were turning out halfwits who knew the lyrics to Elvis Presley songs but could name more members of the Mouseketeers than they could elements of the Periodic Table. A hugely expanded government, of course, was the answer.

NSF director Alan Waterman saw the opening and ran through it like Jimmy Brown barreling into a hole carved out by his Cleveland Browns linemen. The NSF geared up for a major funding boost and an unprecedented level of involvement in the nation's high-school classrooms. The administration's FY 1959 budget request called for a four-year scholarship program that was slanted toward science and for tripling the NSF budget to $140 million, though Democrats blasted Eisenhower for his niggardliness. Education bureaucrats demanded more, more, more!

New offices, new bureaucracies came into being: the Assistant to the President for Science and Technology, the President's Science Advisory Council, the National Aeronautics and Space Administration, and the National Defense Education Act. The President's Science Advisory Committee (PSAC), founded under President Truman in 1951, was a sleepy backwater of outside advisors until *Sputnik* gave it bureaucratic cachet. In 1959, a Federal Council for Science and Technology was also established; its members were government officials. The science bureaucracy would expand further in 1962 with the creation of the Office of Science and Technology, which was, in Jerome B. Wiesner's words, "a focal point for comprehensive science policies in the Executive Branch."[14] Wiesner's own position, that of Special Assistant to the President for Science and Technology, was a product of the panic year of 1957.

As historian of the space age Walter A. McDougall has written, the United States responded to this "first assault on the cosmos" — which had been "launched from the world's first technocratic state" — with "massive state-sponsored complexes for research and development."[15] A nation that had consecrated Big Government and the worship of the state had beaten the U.S. into space, and U.S. leaders were convinced that only by aping the methods of their Soviet rival could the land of the free establish its own position in the skies. The copycatting was not to be limited to science, though science was to be its flashiest battleground.

"The First World War never ended in Russia," writes McDougall. "There alone the methods of state mobilization of scientific and technical talent pioneered in the first total war became a centerpiece of peacetime policy." One major goal of that policy would be to "prove the superiority of social-ism."[16] Ironically, those Americans who urged that the U.S. learn and borrow from the Soviet model were proving the Russians' point for them.

MIT President Dr. James R. Killian was announced as Ike's science advisor on November 7, 1957. Bearing the title Special Assistant to the President for Science and Technology, a full-time position, Killian was directed to assemble a Science Advisory Committee that "was to be positioned at the very summit of government," as he somewhat breathlessly phrased it in his memoir *Sputnik, Scientists, and Eisenhower: A Memoir of the First Special Assistant to the President for Science and Technology* (1977).[17]

Killian was appalled by the "near-hysterical reaction" to the *Sputnik* launches. *Sputnik,* he later wrote, "was being used to support an orgy of technological fantasies and a speed-up in the arms race." These "hard-sell technologists" irritated Eisenhower, too, to no end.[18] The President understood just what was going on — he had spent a lifetime, after all, in the military, and he knew the way that its branches and its defense contractors lobbied for position and funding.

Dr. Killian was no more enthusiastic about spaceships and trips to the moon and planets than was his boss. In a 1960 speech he decried the prospect of "frantically indulging in unnecessary competition and prestige-motivated projects." Instruments, he said, and not man, could more economically — and successfully — explore other worlds. The Apollo program he would later dismiss as a "scientific spectacular," but Killian of MIT knew quite a bit less about stoking public fears and spurring on large-scale public labors than did Kennedy of Harvard.[19]

The education establishment smelled money in the air. As Paul Dickson notes in *Sputnik: The Shock of the Century* (2001), a National Education Association spokesman sighed that "any nation that pays its teachers an annual average of $4,200 cannot expect to be first in putting an earth satellite into space." Dickson quotes NASA archivist Lee Saegesser: "Frankly, it was a field day for certain interest groups. One could get quoted in the papers or invited to testify in front of a Congressional committee just by coming up with some statistic which showed how bad our schools were compared to those in Russia."[20]

"A consensus grew that the nation's educational institutions were largely to blame for this cold war defeat," wrote Barbara Barksdale Clowse in *Brainpower for the Cold War: The Sputnik Crisis and National Defense Education Act of 1958* (1981). The answer: more money. This was the hole that partisans of federal subsidy of education had been looking for. They ran to daylight, or to the Treasury, to be more precise. And the National Defense Education Act of 1958 (NDEA) was *Sputnik*'s baby.

"The Sputnik crisis transformed the politics of federal aid to education; it altered the terms of the debate and temporarily neutralized much of the opposition," says Clowse.[21] The NDEA — notice that "defense" comes before "education" in the title, significantly — was the result. "To those eager to enlarge the federal role in education, this mood of guilt and reform promised to be a flood tide for launching an otherwise grounded policy" of federal aid to education.[22]

Americans were lagging behind the Russians on any number of fronts, or so it was said. For example, "Ten million Russians are studying English, but only 15,000 Americans are studying Russian."[23] One might think that this was an indication that English was the international language of commerce, and Americans might be pleased that our adversaries were aping us instead of us aping them. But that's not how the pundits saw it.

Hyman G. Rickover, the Vice Admiral of the U.S. Navy associated with the nuclear submarine, was an early proponent of "sweeping federal financial assistance to education and imposition of national standards" — a man ahead of his time, it seems, as his desires have come to pass.[24]

Military men were of great use to the educationists as they sought to exploit *Sputnik* and its fallout for pecuniary and legislative gain. The country was deep in the throes of the Cold War; generals and admirals spoke with an authority that, a decade later, as the nation was mired in Vietnam, they would no longer possess. The cost of not heeding their advice, or so the implication went, was nuclear annihilation.

The *Ladies' Home Journal,* joining in the hysteria, published what it called "the most important and concise summary ever written of the challenges facing the United States today." (How's that for overselling an article?) The author: Admiral Hyman G. Rickover. The subject: Education. The solution: More tax dollars. The cost of not taking his advice: Turning the U.S. into a nuclear wasteland.

Consistent with a tradition stretching back through Waldemer Kaempffert, Admiral Rickover envied the Soviet Union its "ruthless technological efficiency which does not appear to be matched by ours."

One might think that it was a virtue of the United States that it lacked the ruthlessness of the USSR, but to Admiral Rickover's mind, the "Western world" could be "lost simply through inefficiency." What was needed on our part was "human effort and sacrifice."[25] (Not, one hopes, "human sacrifice.")

Everything had changed because of the Cold War. Americans would hear the same mantra in the 21st century, after the horrors of 9/11, as indeed they had heard variations on that going back to the first time a politician had schemed to draw power to the central government and away from the private sector. "We no longer have the freedom of choice that used to be an important feature of our way of life," announced Admiral Rickover, a declaration which would have come as a surprise to his commander in chief, President Dwight D. Eisenhower, who had insisted throughout his presidency that American freedoms must not be sacrificed to the struggle with the Soviet Union. "Yet many of us cling to the illusion that it is part of democracy to starve the public sector of the economy and coddle the private sector," he added. His definition of "coddling" the private sector was what others might call permitting free enterprise to run its course. This was no longer an option in Hyman Rickover's America.

"Are we not taking frightful chances when we allow our dislike of government expansion to blind us to the necessity for government-financed and directed development projects?" asked Rickover. These "development projects" now included education, and in a big way. For "Whether we like it or not" — in other words, tough luck, what I'm about to suggest is inevitable — "education is now our first line of defense, and we neglect it at our peril."

Americans had hardly been neglecting education. It's just that for the first couple of centuries of our history education had been the concern of private academies, religious schools, and, later, public schools that were controlled at the local and state level. These schools, asserted Rickover, were inferior to those of totalitarian nations: "Russia's schools and universities outproduce us in highly qualified members of the professions and in masses of young people below the professional-school level who have a solid grounding in the humanities," he said, not offering any evidence.[26] You just had to take his word for it.

Patriotism, as Admiral Rickover was redefining it, meant a willingness to "devote more of our energies to public service." Satisfying consumer needs, as private industries had been doing, verged on unpatriotic, even treasonous. For "Victory cannot be won with more and better washing machines — not when the other side has more and better missiles."[27] (Rickover must be credited with a good line, however, in another polemic. Referring scornfully to the fluff courses that were coming to dominate high schools even in those allegedly Old School days of the 1950s — things like "love and marriage," in this instance — Admiral Rickover observed that "You can learn to make love outside of school in the good old-fashioned ways."[28])

Among the lessons drawn from *Sputnik* and pounded into American heads ad nauseam was that the free enterprise system was inferior to the Soviet command system not only in efficiency and education but also in morality. And so it was open season for moralists and moralizers. The novelist Sloan Wilson, author of *The Man in the Gray Flannel Suit* (1955), informed the readers of *Life* that whereas in America status may be achieved by making money selling widgets and various useless gadgets, "In the Soviet Union... scientists and technicians were the new aristocrats, and the only way to join their ranks was through academic accomplishment." But wait: weren't scientists the likeliest of all Russian professionals to defect to the West, and didn't Soviet authorities surveil them to an extent unrivalled among other professions? Apparently someone forgot to tell the scientists and technicians of their enviable position in the Soviet hierarchy. If only they had known, perhaps they would have been more contented.

Wilson charged that "America's schools... have degenerated into a system for coddling and entertaining the mediocre." He may not have been wrong, but with the benefit of half a century we can say with almost empirical certitude that his solution — massive federal involvement in the schools — was not quite the cure-all he and his allies expected it to be. And like those *Sputnik*-obsessed (or exploiting) critics, Sloan Wilson, who had once been assistant director of the National Citizens Commission for Public Schools — in other words, an employee of an organization formed to lobby for more funding of public schools — played the nuclear Armageddon card.

"[T]imes have changed," he said, using the platitude that is always supposed to end debate and make opponents seem as if they are living in dark caves impenetrable by the light of knowledge. "Space ships and intercontinental missiles are not invented by self-educated men in home workshops. They are developed by teams of highly trained scientists, most of whom must begin (and get so much of) their education in the public schools."

If you don't pump more money into those schools, goes the implied threat, we will not have as many space ships and intercontinental missiles as the Soviets have. And "It goes without saying" — then why does he say it? — "that the outcome of the arms race will depend eventually on our schools and those of the Russians."[29]

In fact the outcome of the arms race, we can now say, did *not* depend on our schools and those of the Russians, or at least not in the sense that Sloan Wilson meant, which was the extent to which these schools were effectively nationalized through federal subsidy. The nation whose schools had the least centralized direction won that arms race. But Wilson's scare tactics, typical of the era, worked in that they helped shape a political climate in which forward-looking patriotism was becoming identified with greater levels of federal expenditure.

This was expressed most baldly by Max Ascoli, editor of the influential Cold War liberal biweekly magazine *The Reporter*. Writing in his magazine in February 1958, Ascoli gasps over "the shocking disparity between the percentage of our national income spent on education and that spent by Soviet Russia."[30] Again, we have the spectacle of a hawkish intellectual admiring the apparently enlightened governmental practices of the Soviet Union. Ascoli calls for a "massive" infusion of federal monies into the U.S. education system, though with the almost supernaturally naïve proviso that "Large-scale Federal financing of education ought to be so sterilized as to make as difficult as possible any attempt at unrestricted control by the Federal government."

Ascoli refers worshipfully to the wisdom of Admiral Rickover and quotes him at length in his disparagement of the tradition of local control of schools. Speaketh the Admiral: "The greatest single obstacle to a renovation of our education comes from the fact that control, financing and direction of education is, in the U.S., in the hands of many thousands of local school boards, whose members seldom qualify as educational experts.... In no other Western country are educational institutions so precariously placed financially, so dependent on local politicians, on the whim of small communities where few have ever had a higher education....The future looks bleak unless in some way Federal assistance can be made acceptable and some sort of national standard can be established to which diploma and degree-giving institutions must conform."

So much for Ascoli's previously expressed concern about federal control. In fact, he worries that without federal control the benighted rubes of the heartland will be unable to compete with those brilliant Russians, products of their centrally controlled schools. The lesson of *Sputnik*, concluded Ascoli, is that "We cannot be modern, streamlined, atom-powered in our industrial and defense structures and remain bucolic and archaic in our educational system."[31] Thus he identifies a decentralized educational system with archaism, and survival in a nuclear age with federal subsidy.

Still, as much as the Scoop Jackson Democrats and *Life* magazine and the National Education Association and Edward Teller tried to whip up hysteria over the world-historical accomplishments of the Soviets, most Americans weren't buying it. A fall 1958 Gallup poll found that 68 percent thought the United States had "a better educational program" than did the USSR, while 19 percent disagreed.[32]

The stakes of these races into space and nationalized education? "National extinction," according to the House Select Committee on Astronautics.[33] The United States was in grave danger of sliding to Number Two in the space race — almost before the race had even begun! The highly publicized failures of U.S. rockets added to the urgency — even though the Russians were experiencing their own (very *un*-publicized) failures. The press had a field day with the fizzling U.S. rockets, calling them *Puffnik, Flopnik, Kaputnik,* and *Stayputnik.* Americans were being scared into a monumental expenditure about which wise scientists were often dubious.

Marvels Tom Wolfe: "It was made clear that NASA could have practically as much as it wanted.... An amazing period of 'budgetless financing' began."[34]

Sputnik, as Walter McDougall writes, "Changed the nature of the Cold War. It made it 'total': — a global conflict in which science, education, housing, and medical care were as much measures of Cold War standing as human rights or bombers. Science was first: the creation of NASA... then the explosive growth of the National Science Foundation, and the National Defense Education Act.... The shock of Sputnik and Soviet claims to social superiority then helped to break down longstanding resistance by Republicans and Southern Democrats to federal involvement in education generally as well as other social areas and began the greatest flood-tide of legislation in American history, a tide that, as Lyndon Johnson recalled, 'all began with space.'"[35]

The classic objections to federal education aid were marshaled by opponents: It was not authorized by the Constitution. Education is a state and local matter; with federal aid would come federal control. But whereas once such objections carried heft, they now seemed as weightless as Americans had been told astronauts would someday be.

"Until 1958, when [advocates of federal aid to education] found a way to link education to national security, their efforts were largely futile," writes Barbara Barksdale Clowse. "As the public reaction to the sputniks grew into a kind of alarm, those who had worked without success to establish greater federal fiscal responsibility for education in the past sensed that an opportunity had been handed to them — literally from the skies." They grabbed the brass ring, or — more accurately — the key to the treasury. As Clowse writes, "many within the administration and in Congress shrewdly began to reassess the proposals they had been mulling to see if they might not fit under the national-security umbrella that had sheltered other legislative areas in the postwar period." President Eisenhower, who in a couple of years would warn of the threat to American liberty posed by a "scientific-technological elite," remarked that "anything you could hook on the defense situation would get by."[36]

Advocates of a national role in education had been waiting for this opportunity for a long time.

Between 1948 and 1962, by the estimate of political scientists Frank J. Munger and Richard F. Fenno in *National Politics and Federal Aid to Education* (1962), "the relevant House and Senate committees [had] conducted hearings whose published record, by conservative estimate, runs to over 10,000 pages and includes more than six million words of testimony" on federal aid to education.[37]

The first congressional consideration of a federal role in education was in 1870, when the House debated a bill providing for national schools in states judged substandard. Numerous proposals had been offered since, some modest in aim and others broad — as broad as the "Americanization bill" of Senator Hoke Smith (D-GA), introduced in wartime America of 1918, which would have created a Department of Education and directed federal funds toward the "Americanization" of pupils, which can be read (benignly) as encouraging the assimilation of foreign-born students or (malignantly) as a totalitarian measure to make docile subjects of students. President Wilson lent his support, but the old American resistance to federal control was strong enough to keep the bill from becoming law.

The next "emergency" after the World War, the Great Depression, spurred passage of a host of government-enhancing measures, but though some relief aid was directed to teachers no comprehensive program of federal assistance to education was enacted. Nor did the Second World War prove conducive to such legislation. The Cold War was the next "emergency," during which "federal educational legislation of some sort [had] been pending almost continuously."[38]

President Eisenhower at first supported a construction program but not more direct federal aid. The usual lobbies played their assigned roles. John Burkhart of the U.S. Chamber of Commerce testified before a House hearing in 1955: "In good times and bad, in war and in peace, whether Treasury surplus or deficit, the NEA has pursued the notion of Federal aid to education with a singlemindedness of purpose that perhaps has never been equaled by any organization in any field over such a long period of time. It has sought to frighten the wits out of the citizenry with its dire predictions of educational catastrophe if our schools are left to the devices of State and local governments. It has sought to woo and win this same citizenry with persuasive pictures of an educational millennium to be achieved under the panacea of Federal aid."[39]

While the National Association of Manufacturers and the American Farm Bureau Federation generally stood with the chamber as skeptics of federal aid, "The most zealous and faithful opponents of federal aid to education have [been]... the so-called women's patriotic organizations, particularly the Daughters of the American Revolution," according to Munger and Fenno. Though often caricatured as super-patriot snobs looking down their noses at anyone whose ancestry cannot be traced to the first-class cabins of the *Mayflower*, the DAR, in its own words from a 1946 resolution, denounced federal education aid because it "would tend toward the further regimentation and centralization of government."[40] Their ancestors would have agreed. And throughout the *Sputnik* era, DAR-type objections could be heard, if faintly, even though the bringers of federal goodies disparaged advocates of local control as benighted provincials.

Stewart McClure, a liberal Senate staffer who as chief clerk of the Labor and Public Welfare Committee was involved in federal education legislation, vented his disapproval of the mossbacks to historian Gareth Davies: "They clung on to some mythical constitutional principle: the last thing that could happen in the United States was for the federal hand to be laid on local education, which belongs to the hands of the school boards and local council of education or whatever they're called – which of course, are all controlled by the Chamber of Commerce [huh???].... Now what all this was supposed to prevent or forestall I never could figure out, but it was a religious faith. They'd get white and scream and wave their hands in the air about the horrible prospects of this vicious, cold hand of federal bureaucracy being laid upon these pristine, splendid local schools that knew better than anyone what needed to be done, and so forth and so forth.... I don't know, it's a real mythology, but it was real and senators and congressmen had to deal with it."[41]

When *Sputnik* orbited, observes Davies in *See Government Grow: Education Politics from Johnson to Reagan* (2007), "supporters of federal aid to education were able to harness national alarm about this development to their cause." The hysteria's "main effect was to legitimize an expansion in the federal role." So "supporters of federal aid wrapped their proposal in the flag and gave it the title National Defense Education Act (NDEA)."

Staffer McClure recalls of his opponents: "We hammered them into the ground. And, of course, if anybody brought up socialism or something like that, the dreadful specter of socialism, we had Edward Teller and the Hydrogen Bomb to clobber them with!"[42] The liberals and the Cold Warriors joined forces, and the archaic old ideals of the American Revolution were clobbered.

Lyndon Baines Johnson, never one to understate matters, declared in opening the Senate Armed Services Committee Subcommittee on Preparedness Investigation in November 1957: "We meet today in the atmosphere of another Pearl Harbor." That's it: Senator: fan panic, act rashly. "We are in a race for survival, and we intend to win that race."

Vannevar Bush, a witness before LBJ's subcommittee, called the lack of federal aid to education "appalling."[43] Edward Teller praised Soviet schools for prizing science education.

Sputnik 2 went up November 3, 1957, carrying a doomed dog named Laika, and the hysteria ratcheted up. The *Sputniks* had opened a wormhole in the space-time-subsidy continuum, and money flowed from it. The AFL-CIO saw the resultant legislation as a jobs program centered on school construction. The NEA saw it as a huge raise for teachers. Cold Warriors viewed it as acceleration in the race against Russia.

Educationists on Capitol Hill wanted federal funds for construction, for a national science academy à la West Point, for college scholarships, and for teacher pay. There were differences of opinion over whether states and localities or the feds should determine where the money was spent, but that the money should be spent was agreed upon.

The NEA gravely announced that the "advances made by the Soviet Union, as shown by the two satellites, may awaken our nation to the importance of stronger school programs." The NEA was demanding the moon and the stars. Senator Wayne Morse of Oregon, later feted as a Vietnam dove, urged the NEA to play the defense card, for "this was the most effective argument with the Senate." NEA lobbyist James McCaskill said, "The bill's best hope is that the Russians will shoot something off."[44]

James Killian, Eisenhower's science advisor, viewed with dismay the "storm of criticism" that education-aid advocates were directing at the American educational system. "Little of this criticism was well informed and thoughtful," he said; "nonetheless it struck close to home."[45] For the purpose of the criticism was to win more monies for the system and to transfer control of that system from its historically decentralized base to a centralized bureaucracy in Washington, DC.

President Eisenhower's special message to the Congress on education of January 27, 1958, was hardly a call to arms, funding-wise. It was clearly the statement of a man who was of two minds on the subject, and who had only arrived at this position after a good deal of conflicted reflection.

"Education best fulfills its high purpose when responsibility for education is kept close to the people themselves – when it is rooted in the home, nurtured in the community," began the President, the purpose of whose address was to offer proposals flying in the face of that opening statement. Reading it today, it is almost as if Eisenhower is trying to talk himself out of what he is about to propose. The tie "linking home and school and community... must be strengthened," he says, and he lists various contributors to that strengthening: parents, school boards, city councils, teachers, principals, state legislators, trustees and faculties of private schools, and others. Conspicuously lacking is any mention of the federal government.

But the feds do enter the picture using their ever-useful passkey: an emergency. Eisenhower concedes, using the language of the scare-tactic but hedging his remarks, obviously uncomfortable with what he is calling for: "Because of the national security interest in the quality and scope of our educational system in the years immediately ahead, however, the Federal government must also undertake to play an emergency role." It's the tried-and-true formula for government spending: National security + emergency = federal dollars.

Eisenhower had played the bureaucratic game at top levels for well over a decade, so he was hardly surprised to report that the director of the National Science Foundation and the Secretary of Health, Education, and Welfare had urged major funding increases and new policy initiatives for... the National Science Foundation and the Department of Health, Education, and Welfare.

The "certain emergency Federal actions" Eisenhower recommended included:

- A fivefold boost for the NSF's science education programs in order to "help supply additional highly competent scientists and engineers vitally needed by the country at this time." The federal government would involve itself in course content, teacher training, and financial assistance to graduate students.
- New programs for the Department of HEW, albeit "for a four-year period only" — wink, wink — which entail grants to the states to encourage testing for science aptitude "at an early stage" in a student's education; grants to the states to subsidize high-school guidance counselors, who presumably will steer promising students into physics and away from such useless subjects as English; college scholarships for "students with good preparation or high aptitude in science or mathematics"; grants to the states to hire science teachers and buy laboratory equipment; financial assistance to potential science teachers in their graduate education; and subsidies to schools to encourage "instruction in foreign languages which are important today but which are not now commonly taught in the United States" — a long and circumlocutory way of saying "Russian."

President Eisenhower, as if trying to reassure himself if no one else, emphasized that "This is a temporary program and should not be considered as a permanent Federal responsibility."[46] But how many major expansions of federal power are ever "temporary"? There would be other emergencies to justify a continued federal role in education, and after a while emergencies would no longer be needed as rationales — this would just be considered something that the federal government does as a matter of course, and no one but a right-wing extremist would even think to object.

Still, the educationists were not impressed. They had the momentum, and they were not about to settle for any half loaves. Rep. Graham Barden (D-NC), chairman of the House Education and Labor Committee, grumbled, "Somebody around [Ike] apparently is of the opinion that all you have to do is to drop a few million dollars in a slot machine, run around behind, and catch the scientists as they fall out." Senator Lister Hill (D-AL), a longtime supporter of federal aid to education, pushed his own bill down the path of nonresistance, saying that the "urgent needs of national defense make early action on this legislation absolutely essential."[47]

The Democrats pressed their advantage. Senator Hill and Rep. Carl Elliott, both Alabama Democrats, called for 40,000 college scholarships annually for six years. It was a mere upping of the ante. No way could a budget-conscious executive such as Eisenhower win this game. Once he had conceded the point on federal involvement, the only question was how high the pot would go. The Democrats called for "the intellectual preeminence of the United States, especially in science and technology."[48] The NEA, they seemed to think, knew the way.

Elliot Richardson, Assistant Secretary of HEW and the epitome of the blueblood Eastern Republican, was the administration's point man on Capitol Hill for the bill. As the legislation wended its way through the Congress, math, science, and engineering were actually *de-emphasized*. Yet the title and the defense gloss of the bill remained, for they were useful selling points.

In the Senate Labor and Public Works Committee, only Strom Thurmond (D-SC) and Barry Goldwater (R-AZ) voted no. Senator Thurmond called out the Cold Warriors, proposing that those who received graduate fellowships under its provisions be required to teach "subjects related to national defense." Thurmond mocked the bill's "unbelievable remoteness from national defense considerations."[49] Keeping with the Cold War theme, the Senate narrowly adopted an amendment by Karl Mundt (R-SD) requiring recipients of loans, fellowships, and scholarships to take a loyalty oath. (The final bill would remove federal scholarships though it kept loans.)

Although at a press conference the president repeated that "the educational process should be carried on in the locality," and despite the fact that on numerous occasions Eisenhower stressed the temporary nature of the program, it is hard to see how a man as skilled at politicking as Eisenhower could ever have believed that the educationists, after supping at the federal trough for four years, would back away without putting up the fiercest resistance.

Whatever his intentions, "Eisenhower was beginning a process of federal involvement that set important precedents for the much more extensive educational programs that Congress would enact in the 1960s."[50]

The NDEA won final passage in the Senate by a vote of 66–15 and in the House by 212–85 and was signed into law by the President on September 2, 1958. The NDEA authorized in excess of $1 billion, which was big money in those days, to be funneled through the Department of Health, Education, and Welfare.

What were the provisions of this National Defense Education Act of 1958 that was going to ensure the intellectual preeminence of the United States? Surely this would be a legislative masterpiece worthy of the sharpest and most sagacious minds the country had to offer.

Finding that "an educational emergency exists and requires action by the federal government," it protested, rather too loudly, in Title I that "FEDERAL

CONTROL OF EDUCATION [IS] PROHIBITED." That throat-clearing out of the way, the NDEA got down to business.

Title II of the act authorized student loan funds at colleges and universities. These funds were allotted to states on a per capita basis depending upon the state's number of enrolled students. Maximum size of a loan was $1,000 annually, or $5,000 in total. Preference was given to students who desire to teach, who are "superior academically," or who display "a superior capacity for or preparation in science, mathematics, engineering, or a modern foreign language."

Title III authorized federal aid to states "for purchase of equipment and as loans to private schools to purchase equipment" to assist in the teaching of science, math, and modern foreign languages (not Latin!).

Title IV created National Defense Fellowships (1,000 in FY '59 and 1,500 in FYs '60–'62) for graduate study, with preference given to students who wish to become "college teachers."

Title V authorized appropriations to secondary schools for "testing to identify abilities and counseling and guidance to encourage students to develop their aptitudes and attend college." The magic words science and math do not appear here.

Title X, Miscellaneous Provisions, included Senator Mundt's loyalty oath, which students had to swear in order to receive funds. The oath, which had passed the Senate narrowly, was later repealed. But it drove home the point that the NDEA was meant first and foremost to win the Cold War, not expand the fund of human knowledge.[51]

In its first year of operation, two-thirds of the nation's 2,000 or so colleges and universities participated in the NDEA. The graduate fellowships were fully utilized and would be boosted from 1,000 to 1,500 the next year. The "emergency undertaking" was here to stay.

Arthur S. Flemming, Eisenhower's Secretary of HEW, spoke of "federal stimulation without federal domination," a somewhat provocative phrase which he used to mean that one can go on the federal dole without becoming a federal dependent.[52] This is also called wishful thinking.

The act further entangled the federal government in higher education, eventually pumping billions of dollars into the system. Its beneficiaries included a University of California at Berkeley math professor named Theodore Kaczynski, later infamous as the Unabomber, as well as the much-derided and eventually junked "New Math." But more than anything, the NDEA was an object lesson in how media-fed hysteria moves, if not mountains, then at least legislatures. A small ball in the air launched by the Russians had altered American education forever.

Upon signing the National Defense Education Act into law on September 2, 1958, President Eisenhower began his brief three-paragraph statement by once again repeating, as if the very act of saying it would make it happen, that the act was "an emergency undertaking to be terminated after four years." He was obviously trying to put this landmark legislation on a path toward a slow fade into the sunset — a slow fade that would never come. "The federal government having done its share," he said, as if wiping his hands of the matter, "the people of the country, working through their local and State governments and through private agencies, must now redouble their efforts toward" the improvement of American education.[53]

The NDEA did not expire, as President Eisenhower had perhaps wanly hoped, but rather was expanded, then absorbed and superseded, by Great Society education programs.

Walter A. McDougall is worth quoting at length on the Eisenhowerian vision, which, despite — or perhaps because of — the President's military background, was suspicious of the rising statist consensus. Eisenhower deplored "a command economy in state-funded research and development... for that meant abandoning the concept of a free society, which develops naturally according to myriad choices on the local level, and replacing it with a central directorate, which charts the path of social progress and orders fabrication of techniques for traversing it. This transition had already occurred in Western Europe (not to mention in the Soviet Union, which was founded on the principle), but it occurred in the United States only in the 1960s — and the catalyst was the space program."[54]

On a parallel track to the NDEA, the President proposed the creation of what became NASA on April 2, 1958; its gestation period was shorter than that of a human infant, let alone a multibillion dollar agency. It was born on October 1, 1958, about a month after the NDEA came into being. The aerospace industry was as fully a child of the federal government as any industry can be. The only market it served was Washington.

The floodgates had opened — literally. At a 1958 hearing of the Defense Appropriations Subcommittee of the House Appropriations Committee, Rep. Daniel Flood, the theatrical Pennsylvania Democrat, went into a fit of ecstatic and rhapsodic appropriating when presented with the testimony of Air Force Assistant Secretary Richard E. Horner. Gushed Flood: "If this man means what he says [about the possibility of man landing on the Moon] and if he knows what he is talking about — and I think he does — then this committee should not wait five minutes more today. We should give him all the money and all the hardware and all the people he wants, regardless of what anybody else says or wants, and tell him to go on top of some hill and do it without any question."[55]

Flood spoke for many. NASA and the military should have whatever they want, as much as they want, and they should be allowed to go off and do it without pesky civilian oversight. This is inconsistent with American constitutional and republican principles, but when men fall under the spell of the space race, reason seems to take its leave.

When Yuri Gagarin, along for the ride on the *Vostok 1*, achieved the first manned orbital flight of a spacecraft on April 12, 1961, the panic in the White House would go into overdrive. On April 20 of that same year, President Kennedy asked Vice President Johnson to find aerospace contests at which the U.S. could best the Soviet Union. Was there "any other space program which promises dramatic results in which we could win?" asked the President.[56]

Well, yes, as it happened, NASA officials and prominent members of the astronautical community thought we could beat the Russians to the Moon. And so on May 25, 1961, President Kennedy told a joint session of Congress: "I believe this nation should commit itself to achieving the goal, before this decade is out, of landing a man on the moon and returning him safely to the earth. No single space project in this period will be more impressive to mankind or more important for the long-range exploration of space; and none will be so difficult or expensive to accomplish."[57]

It sure was expensive.

Carl Sagan wrote in wonderment that Kennedy science advisor Jerome Wiesner told him that he had a deal with President Kennedy: if JFK didn't pretend Apollo was about science, then Wiesner would not object to this obviously political program: "Here was the United States, ahead of the Soviet Union in virtually every area of technology — the world's economic, military, and, on occasion, even moral leader — and Indonesia would go Communist because Yuri Gagarin beat John Glenn to Earth orbit? What's so special about space technology?"[58]

The space program is the very flashy and panic-borne program to which we turn our attention in the next chapter.

Notes

1. Robert A. Divine, *The Sputnik Challenge* (New York: Oxford University Press, 1993), p. xiv.
2. Tom Wolfe, *The Right Stuff* (New York: Bantam, 1980), pp. 56–57.

3. Golf rounds courtesy of Eisenhower Presidential Library, Abilene, Kansas.
4. Quoted in *The Politics of American Science: 1939 to the Present*, p. 209.
5. Quoted in Divine, *The Sputnik Challenge*, p. xv.
6. Paul Dickson, *Sputnik: The Shock of the Century* (New York: Walker & Company, 2001), p. 112.
7. Quoted in ibid., pp. 120, 117, 204, 118.
8. Quoted in Divine, *The Sputnik Challenge*, p. 98.
9. Ibid., p. xv.
10. "Defense: Orderly Formula," *Time* (October 28, 1957).
11. Quoted in Divine, *The Sputnik Challenge*, pp. xiv–xv.
12. Quoted in Dickson *Sputnik: The Shock of the Century*, p. 225.
13. Quoted in Divine, *The Sputnik Challenge*, pp. viii, xvi, 7, 15, 16. Divine provides an excellent survey of contemporary responses to *Sputnik*.
14. Wiesner, *Where Science and Politics Meet*, p. 66.
15. McDougall, "Technocracy and Statecraft in the Space Age — Toward the History of a Saltation," pp. 1014, 1010.
16. Ibid., p. 1015.
17. James R. Killian, *Sputnik, Scientists, and Eisenhower: A Memoir of the First Special Assistant to the President for Science and Technology* (Cambridge, MA: MIT Press, 1977), p. xv.
18. Ibid., pp. xvii, 238–39.
19. Ibid., pp. 143.
20. Quoted in Dickson, *Sputnik: The Shock of the Century*, p. 226.
21. Barbara Barksdale Clowse, *Brainpower for the Cold War: The Sputnik Crisis and National Defense Education Act of 1958* (Westport, CT: Greenwood Press, 1981), pp. 3–4.
22. Ibid., pp. 12–13.
23. Sloan Wilson, "It's Time to Close Our Carnival," *Life* (March 24, 1958), p. 36.
24. Clowse, *Brainpower for the Cold War: The Sputnik Crisis and National Defense Education Act of 1958*, p. 35.
25. H.G. Rickover, "Your Child's Future Depends on Education," *Ladies' Home Journal* (October 1960), p. 98.
26. Ibid., p. 100.
27. Ibid., p. 102.
28. Wilson, "It's Time to Close Our Carnival," p. 36.
29. Ibid., p. 37.
30. Max Ascoli, "Our Cut-Rate Education," *The Reporter* (February 20, 1958), p. 8.
31. Ibid., pp. 8–9.
32. Divine, *The Sputnik Challenge*, p. 186.
33. Wolfe, *The Right Stuff*, p. 58.
34. Ibid., p. 229.
35. McDougall, "Technocracy and Statecraft in the Space Age — Toward the History of a Saltation," p. 1030.
36. Clowse, *Brainpower for the Cold War: The Sputnik Crisis and National Defense Education Act of 1958*, pp. 44, 53.
37. Frank J. Munger and Richard F. Fenno Jr., *National Politics and Federal Aid to Education* (Syracuse, NY: Syracuse University Press, 1962), p. 2.

38. Ibid., pp. 5, 8.
39. Quoted in ibid., p. 20.
40. Ibid., pp. 29, 31.
41. Gareth Davies, *See Government Grow: Education Politics from Johnson to Reagan* (Lawrence, KS: University Press of Kansas, 2007), p. 11.
42. Ibid., pp. 12–14.
43. Quoted in Clowse, *Brainpower for the Cold War: The Sputnik Crisis and National Defense Education Act of 1958*, pp. 59–60.
44. Quoted in ibid., pp. 73–74, 77.
45. Killian, *Sputnik, Scientists, and Eisenhower: A Memoir of the First Special Assistant to the President for Science and Technology*, p. 191.
46. Dwight D. Eisenhower, "Special Message to the Congress on Education," January 27, 1958, The American Presidency Project, http://www.presidency.ucsb.edu/ws.
47. Quoted in Clowse, *Brainpower for the Cold War: The Sputnik Crisis and National Defense Education Act of 1958*, pp. 79, 87.
48. Divine, *The Sputnik Challenge*, p. 157.
49. Quoted in Clowse, *Brainpower for the Cold War: The Sputnik Crisis and National Defense Education Act of 1958*, pp. 125–26.
50. Divine, *The Sputnik Challenge*, pp. 162, 166.
51. National Defense Education Act of 1958, U.S. Statutes P.L. 85–864.
52. Quoted in *The Politics of American Science: 1939 to the Present*, p. 210.
53. Dwight D. Eisenhower, "Statement by the President Upon Signing the National Defense Education Act," September 2, 1958, The American Presidency Project, http://www.presidency.ucsb.edu/ws.
54. McDougall, "Technocracy and Statecraft in the Space Age – Toward the History of a Saltation," p. 1025.
55. Quoted in Carl Sagan, *Pale Blue Dot: A Vision of the Human Future in Space* (New York: Random House, 1994), p. 169.
56. Quoted in Howard E. McCurdy, *The Space Station Decision: Incremental Politics and Technological Choice* (Baltimore, MD: Johns Hopkins University Press, 1990), p. 13.
57. John F. Kennedy, "Special Message to Congress on Urgent National Needs," May 25, 1961, http://www.presidency.ucsb.edu/ws.
58. Sagan, *Pale Blue Dot: A Vision of the Human Future in Space*, p. 161.

Chapter 5

To Mars! (But Why?)

The space race that was one of the central and thrilling narratives of the 1960s would not have been possible without the stimulus that has so often been the goad in federal science policy: Fear. As Stephen Baxter wrote in *Spaceflight*, the journal of the British Interplanetary Society, "Throughout its history, American government has been at its most effective in reacting to threats, perceived or real."[1] By effective, of course, he means expansive — and expensive.

Tom Wolfe's book on the Mercury astronauts, *The Right Stuff*, entertainingly conveys the combustible mixture of excitement, the thrill of discovery, and the exploitation of public fears that fueled the early space program. Although NASA would far outpace the Soviet aeronautical effort, the first few years of manned space flight were a kind of U.S.–Soviet duel, or an extended game of one-upsmanship. The Soviet Union launched *Sputnik 1*; the U.S. launched *Explorer 1*. The Soviets sent dogs into space, most famously Laika; the U.S. sent up chimpanzees. The Soviets sent Yuri Gagarin up in *Vostok 1*; the U.S. responded with Alan Shepard in *Freedom 7*.

After Gagarin became the first human being to enter orbit on April 12, 1961, President Kennedy asked Vice President Johnson, "Do we have a chance of beating the Soviets by putting a laboratory in space, or by a trip around the Moon, or by a rocket to go to the Moon and back with a man?"[2]

It was a competition, pure and simple. War by peaceful means. You know who won that competition. Whether it was worth the $25 billion cost (that's

J.T. Bennett, *The Doomsday Lobby: Hype and Panic from Sputniks, Martians, and Marauding Meteors*, DOI 10.1007/978-1-4419-6685-8_5,
© Springer Science+Business Media, LLC 2010

in 1969 dollars; in 2010 dollars the total is more than six times as much) is another question.

Although the Moon was NASA's target in the 1960s, "NASA explored the technical feasibility of a Mars mission in as many as 60 study contracts between 1961 and 1968."[3] And, in fact, plausible, if outrageously expensive, plans for sending human beings to Mars had been floated since the 1950s.

The Red Planet — "Mars, the Bringer of War," in composer Gustav Holst's classic work *The Planets* — has exerted a hold on the imaginations of men and women unrivalled by any other celestial body. The aforementioned Percival Lowell, the blueblood eccentric who founded the Lowell Observatory in Flagstaff, devoted himself to proving the existence of life on Mars in the late 19th century. Working from a mistranslation of the Italian word *canali*, which means "channels," Lowell picked up on the work of the Italian astronomer Giovanni Schiaparelli, who in 1877 had claimed to see *canali* on the surface of Mars. (That year was a milestone in Mars studies, as Asaph Hall, working at the Naval Observatory, discovered the two tiny moons of Mars, Phobos and Deimos. John Quincy Adams's lighthouse had finally made its mark.)

To Percival Lowell, the *canali* were canals, or evidence of intelligent life. They were man-made — Martian-made, that is — and they suggested a highly evolved life form that had undertaken to bring water from the Martian poles to the dry areas closer to its equator. Lowell's ideas were fanciful, cut from whole cloth, and had not the slightest resemblance to Martian climatology, but he was doing this work on his own dime, and surely the endower of a major observatory has the right to pursue his own lines of inquiry, however odd they may be.

Lowell's books of Mars were a sensation. And they were not the only popular works to imagine life — sometimes quite malevolent life — on our planetary neighbor. The popular astronomer Simon Newcomb, a kind of Carl Sagan of the early 20th century, declared, "There appears to be life on the planet Mars. A few years ago this statement was commonly regarded as fantastic. Now it is commonly accepted."[4] The English science fiction novelist H.G. Wells had an enormous success with *War of the Worlds* (1898), in which the Martian race launches an invasion of Earth that is brutal and terrifying and merciless. Edgar Rice Burroughs, the American fantasist whose most enduring creation was *Tarzan of the Apes* (1912), also began in 1912 his Barsoom series of novels about derring-do on the Red Planet. The first entry in the series was *A Princess of Mars*, a novel which held out the alluring possibility that Mars was populated by scantily clad and beautiful women who had a thing for brave Earthmen. American teenage boys found the Barsoom books irresistible. ("Barsoom" is what Martians call their planet, in Burroughs's fictive world.)

Among those entranced by the Burroughs novels were many of the Americans who grew up to become the rocketeers, astronomers, and aerospace engineers who put men on the Moon and dreamed of putting men on Barsoom. As NASA librarian Annie Platoff wrote in her informative and valuable manuscript "Eyes on the Red Planet: Human Mars Mission Planning, 1952–1970," the "stories of travel to Mars significantly influenced early human Mars mission planners." The influence crossed cultures, even oceans. "Rocket pioneer Werner von Braun noted that, just as many American scientists had been influenced by a childhood fascination with Burroughs' Martian novels, many of the German rocket scientists had, as children, 'buried themselves in the pages of *Auf Zwei Planeten'* " — that is, *Two Planets*, an 1897 novel of a Martian civilization by Kurd Lasswitz.[5] As in the novels of Burroughs, life on Mars was depicted as much more exciting than the dull old doings on Earth. Mars was where the action was. The boys become men who read those novels were inspired to want to help earthlings leave their planet and visit the next farthest from the Sun.

Werner von Braun, the German scientist who led the team that produced the Nazi V-2 rocket, was, writes Platoff, "[w]ithout a doubt, the most influential figure in the history of human Mars mission planning."

Von Braun was hardly an ardent Nazi; late in the war he was imprisoned by the Gestapo on suspicion of a lack of sympathy for the German war effort. His interest was in civilian, and not military, rocketry, though the applications of his research were military — destructively so. In 1945, von Braun and other leading figures of German rocketry defected to the Americans; the transfer of these scientists from German to U.S. auspices was called Operation Paperclip. And while his early work for the U.S. had the same military-over-civilian emphasis he had so disliked in Germany, eventually von Braun became the most public advocate of a space program with a civilian appeal — even if its core, and its justification, remained military.

In 1953, von Braun published *The Mars Project*, in which he sought to put to rest dime-novel fantasies about eccentric rich men and ingenious but penniless science whizzes teaming up to build rocket ships out in the meadow. "[T]rue space travel," he lectured confidently, several years before anyone had actually achieved true space travel, "cannot be attained by any back-yard inventor, no matter how ingenious he might be. It can only be achieved by the coordinated might of scientists, technicians, and organizers belonging to nearly every branch of modern science and industry." The phrase "coordinated might" was somewhat frightening coming from a man who had, as he sometimes ruefully said, aimed at the Moon and hit London. But von Braun was a true believer in big government, and "interplanetary exploration," he announced, "must be done on a grand scale."[6]

As Annie Platoff explains, von Braun's "grand scale" had three score and ten astronauts being ferried on ten craft from Earth to Mars on 260-day voyages to and fro, with a Martian stay of 15 months. Von Braun's Mars Project got a massive publicity sendoff with a series of cover stories in the three-million-circulation, general-interest magazine *Collier's*. Illustrated by the incomparable space artist Chesley Bonestall, the *Collier's* articles led to a three-episode television series produced by the Walt Disney studio. The impossible was coming to seem inevitable. "It didn't matter," writes historian of manned spaceflight Andrew Chaikin, "that von Braun's grandiose plans bore scant relation to reality... because their psychic impact was undeniable."[7] Humans were going to Mars. The pictures were already in *Collier's*, for Pete's sake.

By 1965, von Braun, then director of NASA's Marshall Space Flight Center, was telling readers of *Astronautics & Aeronautics* that eight-man missions to Mars — "even the establishment of a semipermanent" manned base — using Apollo-style hardware would be possible by 1985.[8] Von Braun's proposed Mars stopover would take 456 days from takeoff to return home, with twenty of those days spent on the Martian surface.

Werner von Braun did not always show the best political judgment. In 1937 he had joined the National Socialist German Workers Party — the Nazis — and while he claims to have done so under official pressure, the act does tend to cast a shadow over his subsequent judgments. He certainly misunderstood the nature of a democratic republic. In his 1965 article setting forth his proposal for a Mars mission, he suggests that a semi-permanent base might be established on the very first mission. The "easiest way of doing it," he writes, "would be to send the first expedition to Mars without a complete return capability. In other words, give this crew all the weight that they would otherwise need in a return flight for extended survival on the planet, and tell them that, Congress willing, we would hope to raise enough money the following year to pick them up again."[9]

Oh sure: *that* would go over like a lead zeppelin. Another name for what von Braun proposes is blackmail, or ransom: Give us X billion dollars for a Mars trip or the eight heroic spacemen die. Is it even remotely conceivable that any elected government anywhere, at any time, would do such a thing?

Von Braun, in answering the question, "Why do we have to go to Mars?" responds with a highly speculative, if at the time fashionable, imperative. "I have not heard of a good scheme to legislate away the population explosion as yet, so maybe one of these days we'll have to provide room for new settlers on other planets."[10]

In other words, ship the excess population to Mars, or Venus, or whatever planet we might terraform so as to be hospitable to Earth creatures. As critics noted, this would be a practical impossibility: if the world population increases each day by over 200,000, that would mean that in order to achieve a static population spaceships would need to expel — or rather export — over 200,000 persons per day, every day. Even with a spaceship-building program that is exponentially greater than the New Deal multiplied by the Obama stimulus package, it's hard to see how 200,000 people could be transported from the planet — involuntarily? — every single day. Or maybe the U.S. could undertake a crash program to develop a *Star Trek*-like transporter that would simply dematerialize the traveler on earth and rematerialize him on Mars. That technology, we can safely say, is several years away.

As space reformer Robert Zubrin later wrote, to von Braun, "a manned interplanetary mission was part and parcel of a hardware fabricator's wildest dreams."[11] It was all machinery, no exploration; all hardware, no science; all dough, not much show. It was a year-plus in flight and not even three weeks on the surface. It was a subsidy beyond the aeronautical industry's wildest dreams.

NASA kept Mars at arm's length. In its *Long Range Plan of the National Aeronautics and Space Administration* (1959), the fledgling bureaucracy did not set a timetable for a human mission to Mars, allowing only that it might happen "beyond 1970." In keeping with President Eisenhower's moderate space skepticism, the President's Science Advisory Council, in its December 16, 1960, Report of the Ad-Hoc Panel on Man-In-Space, said that "manned trips to the vicinity of Venus or Mars are not yet foreseeable."[12]

But then came John F. Kennedy, and many things that were once unforeseeable were now not only glimpsed on the horizon, but planned for and paid for as well.

NASA let three contracts in May 1962 for the study of manned flights to Mars and Venus. Aptly — or ominously — dubbed "Project EMPIRE," its contractors (Ford's Aeronautic Division, General Dynamics/Astronautics, and Lockheed Missiles and Space Company) submitted reports detailing how such missions might be undertaken by 1975. In 1963, two more mission-to-Mars studies were commissioned, this time from TRW Space Technology

Laboratories and North American Aviation. The envisioned trips were long
on aeronautical virtuosity and short on planetary science: typically, the travel
time was well over a year and the stay on Mars was as short as a single week.
It was an awful lot of money and sweat and tears and danger just to pick up
a few rocks and bid the Red Planet adieu.[13]

The early 1960s were an age of frenetic and wondrous activity at NASA.
More Mars studies followed, and these "confirmed the belief of many at
NASA that human missions to Mars were feasible without significant technological developments beyond those already under way at the time," writes
Platoff.[14] By the early 1980s at the latest, went the conventional wisdom at
NASA, spacecraft would hurtle toward Mars, and upon arrival men would
disembark, planting the flag of the United States in the coppery soil. It
seemed so... inevitable. Apollo would take us to the Moon, and a subsequent
voyage to Mars would be as natural as taking one more step up a high
ladder.

A National Academy of Sciences committee, the Space Science Board,
recommended in 1965 that exploration of Mars take precedence over the
Moon, as a fuller knowledge of matters Martian would better help us understand "the origin of the solar system, the origin of life, and the understanding of the Earth." But in retrospect, these heady years of the mid-1960s were
the peak of let's-go-to-Mars ballyhoo. In August 1965, as Annie Platoff
notes, Donald F. Hornig, President Lyndon B. Johnson's science advisor,
threw cold water on the Mars ardor before a Senate committee. Hornig
speculated that a manned trip to Mars could cost five times as much as the
entire Apollo program, or a then-astronomical $100 billion. Other estimates
put the cost closer to $50 billion, still a gigantic number. Given the admin-
istration's priorities in Asia and in American cities, "a number of national
objectives... seem more urgent."[15] When an LBJ advisor worries that the
price tag of a federal program is just too darned high, you know you've got
a major boondoggle on your hands.

The NASA budget began falling. It went from $5.93 billion in the peak
year of 1966, which was an amazing 5.5 percent of the total federal budget,
to $5.43 billion in 1967 (3.1 percent of the budget), $4.72 billion (2.4
percent) in 1968, $4.25 billion (2.1 percent) in the moon landing year of
1969, and ever downward. Lyndon B. Johnson's guns and butter orgy was
leaving limited budgetary space for the rocketmen.

In 1967, NASA was barred from spending money on Mars studies. The
last real window of opportunity in the 20th century for a grand trek to Mars
came with the election of Richard M. Nixon to the presidency in 1968 and

the contemporaneous elevation of Thomas O. Paine, deputy administrator of NASA, to the directorship.

President Nixon, in the optimistic first year of his presidency, selected Vice President Spiro Agnew to chair the Space Task Group (STG), which was charged with determining what new challenges NASA should take on after the conquest of the Moon. The first NASA administrator under Nixon was the patriotically named Paine, a metallurgist who had spent his career with General Electric as a researcher and manager in its laboratories. The preference for private over public endowment of science that was advanced with such eloquence and conviction by GE's Frank Jewett seems not to have rubbed off on Dr. Paine, however. His year and a half (March 21, 1969–September 15, 1970) as NASA administrator coincided with the agency's two greatest moments — the landing of *Apollo 11* upon the Moon and the successful return to Earth of the damaged *Apollo 13* craft. Paine was a "visionary" (he later called himself a "visionary prophet"[16]) who, in science-fiction author Stephen Baxter's assessment, failed to understand that the United States could not simultaneously expend blood and treasure in Vietnam and brains and treasure on an ever-expanding NASA whose sights were to shift from the Moon to Mars.[17] Paine said in August 1969, a month after Neil Armstrong's famous step, that the U.S. space program could put a man on Mars by 1982; in Annie Platoff's words, "such a mission would only require the will of the nation to do it."[18] As well as an open purse — but that wasn't Thomas O. Paine's problem.

Spiro Agnew was the corrupt face of a manned expedition to Mars, an "Apollo of the 70s"[19] plan that stood to squander even more public funds than had the mission to the Moon. Vice President Agnew, who would resign in disgrace just over four years later, said in the rosy shadow of the successful *Apollo 11* landing that the government "should articulate a simple, ambitious, optimistic goal of a manned flight to Mars by the end of this century."[20] He was, he said, "all out for space."[21] Agnew expected to find audiences full of Buck Rogers fans and readers of Edgar Rice Burroughs who couldn't wait to find voluptuous seductresses waiting for us on Mars, but he found nothing of the sort. He was actually booed when plumping for a Martian expedition.

Early indications were that Agnew's STG was going to give Earth's Martians — who were not little green men but instead were suspiciously concentrated in the aeronautical-industrial complex — everything they wanted and more. At the task force's second meeting, he asked those gathered, "What about a manned mission to Mars?"[22] It was a rhetorical question.

The head of the Mars mission group within the STG was none other than Werner von Braun. The fix, it would appear, was in.

An April 1969 draft report stated forthrightly, "We recommend that the US begin preparing for a manned expedition to Mars at an early date." (As Baxter points out, this would be weakened in the final report to "Manned expeditions to Mars could begin as early as 1981."[23] The passive voice was a concession of defeat.)

NASA and the aerospace industry were all for a trip to Mars, which meant billions in contracts for the latter and a new raison d'être for the former. On the other hand, reaction from scientists without a direct financial or career stake in a Mars mission was muted, if not openly hostile. Eugene Shoemaker of California Institute of Technology told the *Washington Post* in August 1969 that "it is premature to go to Mars, and the day when we should go there is a long way off." His Cal Tech colleague Bruce Murray added, "Mars isn't going to go away. We have no need to be in a rush to get there." Continued Murray, "The need to land men on Mars just doesn't exist. We're ahead of the Soviets in almost every aspect of space. There is no justification for it at all."[24]

The justification, NASA's Paine told readers of the *New York Times* in July 1969, as *Apollo 11* was headed to the Moon, was that colonizing other planets offered a way of solving social problems at home. "No longer must all of humanity's hopes and fears reside on our shrinking home planet Earth," he said, making it sound as if our globe were getting smaller by the minute, and that surplus people would fall off into the void of space if we didn't hurry up and start sending the teeming billions to nearby planets.[25] As with von Braun, Paine played upon the burgeoning fears of overpopulation and mass starvation to stimulate support for a Martian trip. But those fears lacked the urgency of the Cold War anxieties that gave us a federal role in education, for instance. Russian missiles seemed more real than future famine.

Agnew's Space Task Group laid out five Mars options. The two most interesting — and radical — were provided for "comparison purposes" only. The first, and most ambitious, and most expensive, and the stuff of Werner von Braun's dreams, called for a manned mission to Mars to leave Earth on November 12, 1981, with two ships carrying six men each. A space station and space shuttle would act as waystation and taxi on the voyage. The 12 astronauts would reach Mars in August 1982. A Mars Exploration Module would ferry astronauts from the mother ship to the Red Planet. Their visit would last 30 days, and they would arrive back on Earth on August 14, 1983. By 1989, Mars would be playing host to a 50-person base. This crash program was included in order to make the others seem reasonable. The second, third, and fourth options utilized a space shuttle and space station, but they did not call for Americans to touch down on Mars until 1984, 1986, and indefinitely, respectively. The respective costs were $9 billion annually, $8 billion annually, and $5 billion annually. The fifth option, which the STG considered unreasonable, even wacky, was to "end the human space program after the completion of the Apollo Program." This was the option consistent with fiscal tight-fistedness and a strict construction of the Constitution, but as the nation approached its bicentennial, such casts of mind were considered hopelessly old-fashioned, not even worthy of serious discussion. They were as ridiculous as the ideas of Frank Jewett.

Nixon's budget director, Richard P. Mayo, tore into the Space Task Group's report. He "suggested the cost figures provided by the STG had seriously underestimated the cost of the future space program" and he wondered why more attention had not been given to un-manned alternatives.[26] The space boys had tried to launch one over the heads of the budget hawks, but the hawks flew higher than anyone had expected. White House aides denounced "NASA's empire-building" in internal memos. Top aide John Ehrlichman was a known anti-Mars-mission partisan.

Nixon, interestingly, was less Mars-crazy than Agnew. In fact, as Stephen Baxter writes, "Nixon, by contrast to Paine, seemed to understand the anti-technocratic mood of his day, and also how technocracy was in opposition to America's older Jeffersonian tradition of local politics and democratic responsiveness."[27] Nixon could also read polls. His advisor Peter Flanigan had sent him a memo pointing to a *Newsweek* survey in the fall of 1969 showing that 56 percent of Americans favored cutting the space budget while only 10 percent supported an increase.[28]

Despite the propaganda barrage from the networks, the newsweeklies, and seemingly all of the mainstream media, "NASA was constantly faced with strong public and private opposition to manned spaceflight," writes Baxter.[29]

The Gallup poll of August 1969 found that 53 percent opposed a manned mission to Mars and 39 percent favored it — and this at the peak of Moon-mania. The rocket-man craze that had carried Americans to the Moon was fading. Watergate, Vietnam, and the Great Society had gone sour, and taxes were running amok — the political climate of the early-mid 1970s would be markedly different from the height of the JFK–LBJ–Nixon years. Out in California, a cranky old patriot named Howard Jarvis was leading a tax revolt against the confiscatory real-estate and property taxes that were beggaring the middle class in the Golden State. In Washington, DC, Senator

Frank Church's (D-ID) Select Committee to Study Governmental Operations with Respect to Intelligence Activities was uncovering a myriad of illegal and unconstitutional acts by U.S. intelligence agencies, among them the FBI's and CIA's wide-scale opening of mail to American citizens and the Keystone Kops-like efforts to employ the Mafia to kill Cuban President Fidel Castro. Reform, even radical change, was in the air. Americans no longer trusted the institutions of Big Government. And the space program, for better or worse, was very much part of the military-industrial complex.

The "public and political reaction" to the Space Task Group "was swift and negative."[30] Anti-Mars protesters actually greeted Nixon in August 1969, as he dined with the *Apollo 11* astronauts at a state dinner in Los Angeles. They held signs saying, indelicately, "Fuck Mars."[31] Pollsters had found that support for NASA dropped sharply once Neil Armstrong's boot touched the Moon. Apollo was seen not as some transcendent and ongoing scientific achievement but rather as a race against the enemy: the Russians. We won, we planted the flag, and now it was time to stay home and tend to domestic matters. With each subsequent Apollo mission public support for the space program dropped ever more. In fact, as Sylvia K. Kraemer, director of the Special Studies Division of the National Aeronautics and Space Administration, wrote, "The proportion of Americans *opposed* to more government expenditures in space from 1965 to 1975 increased from one-third to one-half of all adult Americans."

These opponents of the space program were likelier to be drawn from the ranks of women, minorities, and "the less-educated in non-salaried positions."[32] The poor, in other words, or at least the less well-off. These demographic groups tend to generally support greater government spending, yet space exploration is an exception — perhaps because they perceive that this is, as a liberal congressman once charged, a jobs program for engineers and science Ph.Ds. California liberal Democratic Congressman Leo Ryan, later to win dubious fame as a victim of the Jim Jones cult in Guyana, delivered a stinging rebuke to a Mars mission in a House speech in which he lamented that such a voyage would take money from "the unfulfilled human needs on Earth."[33]

For the most part, these liberal Democratic critics of a Mars mission (and even the Apollo voyages) shared the technocratic bias of the space racers — it is just that they wished the federal government to turn its engineering attentions to such domestic problems as housing, urban renewal, medicine, and education. They did not object to an all-wise central state making decisions that earlier generations believed were best made at the local level, by people with immediate experience of the problem. To most (but not all) of the liberal critics, HEW was as sacred a cow as NASA was to the Mars crowd.

Just how unenthusiastic Americans have been about the space program is evident in just about any public opinion poll whose numbers have not been cooked by tendentious phrasing. Jon D. Miller, polling for the International Center for the Advancement of Scientific Literacy, reported that between 1979 and 1990, only 8–10 percent of Americans were part of the "attentive public for space exploration." These "attentives" were hardly Stephen Hawking clones. A majority believed that UFOs were spaceships containing aliens, and upwards of one-third were unaware that the Earth revolves around the Sun in one year's time.[34] It seems that perhaps the science education that was the alleged beneficiary of the fallout from the *Sputnik* hysteria has not been doing the job. Is it time to go back to the drawing board?

On March 7, 1970, Nixon the pragmatist announced that he had chosen the third of the three "realistic" options presented by the Space Task Group. He refused to attach any date or timetable or make a pledge toward a Mars landing. "Essentially, Nixon had put an end to the speculation of the future of human spaceflight," observes Platoff.[35] Thomas Paine quit NASA in September 1970 and went back to GE.

Apollo was "an end in itself, a system designed to place two men on the Moon for three days, and it achieved precisely that." It did not lead systemically or even logically to a Mars mission; it was self-contained. It was also a reaction to the Soviet space program, and the hysteria over early Russian orbital missions had greatly contributed to Apollo's budget. By 1969, however, it was clear to all but the most paranoid that Russia was not even close to a Moon mission, let alone a trip to plant the Red flag on the Red planet.

Still, Nixon did not axe the space program. His deputy director of the Office of Management and Budget, Caspar Weinberger, who styled himself "Cap the Knife," pled for NASA in 1971. Weinberger argued that trimming NASA's budget would reveal "that our best years are behind us, that we are turning inwards, reducing our defense commitments, and voluntarily starting to give up our superpower status and our desire to maintain world superiority." Nixon bought it – and NASA was saved from the perils of fiscal conservatism. "World superiority" saved NASA's bacon.

A Mars mission, some diehards hoped, had the potential to recharge NASA's batteries, to give it a new air of daring and romance. By Nixon's final year in office, the space program represented about 100,000 jobs, or one-fourth the total at its highest point.[36] It badly needed a boost.

The Soviet Union was the first to attempt an unmanned journey to Mars. A series of launches failed, whether on the pad or en route to the planet: *Marsnik 1* and *Marsnik 2*, *Sputnik 22* and *24*, *Mars 1*, and *Zond 1964 A*. The American *Mariner 4* was the first spacecraft to make a successful surveillance trip to Mars, 230 days in length, though it "quashed once and for all the Lowellian vision of the Red Planet, revealing a barren, cratered surface, more Moon-like than Barsoom-like."[37] Mars appeared desolate, devoid of life, devoid even of traces of former life. There was no advanced civilization on Mars, no giant spiders, no intelligent tentacled beings, no gorgeous princesses waiting to be rescued by daring Earth astronauts.

The Soviets, meanwhile, finally achieved a success with *Mars 3* in 1971, but this was overshadowed by the *Mariner 9*, launched in November 1971, and the *Viking* craft sent out by the United States in the mid-1970s. The photos returned from space were duller than anything Wells or Burroughs ever dreamed up.

On the eve of *Mariner*'s successful voyage to Mars, five men who were as closely identified with Mars in the public imagination as any earthling then alive gathered at California Institute of Technology in Pasadena — whose Jet Propulsion Laboratory had sent the ship hurtling Mars-ward — to hold a panel discussion on the lofty topic of "Mars and the Mind of Man." The quintet included science fiction novelists Ray Bradbury (*The Martian Chronicles*) and Arthur C. Clarke (*2001: A Space Odyssey*), astronomer Carl Sagan, Cal Tech planetary scientist Bruce Murray, and, as moderator, the *New York Times* science writer Walter Sullivan.

Reading the transcript, Bradbury is the poet of the bunch, Clarke is Delphic, and the ever quotable Sagan, of course, gets off the best lines. For instance, of the geometrically shaped canals Percival Lowell had seen at his eyepiece four score years earlier, Sagan quipped, "There is no question that the straightness of the lines is due to intelligence. The only question concerns which side of the telescope the intelligence is on."[38] A good line, but really, there is no doubt which side the intelligence is on — or the way in which that intelligence misled its possessor into seeing things that were not there.

The wish being father to the thought — and the ways in which multibillion-dollar Big Science projects can grow from "wishful thinking" and misleading lines of thought — was the subject of the self-proclaimed "black hat" of the gathering, Bruce Murray, the brilliant MIT-trained Cal Tech geologist and planetary scientist who took it upon himself, rather bravely, to tell the assembled Mars fans that "we have erred in being too optimistic" about even the remote possibility of life on Mars and that this unfounded optimism had had, and would have, deleterious consequences for the space program.

Just as in a later generation, the paranormal-friendly FBI agent Fox Mulder of television's *The X-Files* would take as his motto "I Want to Believe," so Bruce Murray charged the Mars crowd with being "captive to Edgar Rice Burroughs and Lowell." This captivity was not a mere matter of whimsical daydreaming. "Mars somehow has extended and endured beyond the realm of science to so grab hold of man's emotions and thoughts that it has actually distorted scientific opinion about it," bemoaned Murray. "[M]an as a human species has been guilty of wishful thinking collectively. We *want* Mars to be like the Earth. There is a very deep-seated desire to find another place where we can make another start that somehow could be habitable." We want to believe — or at least scientists who envision healthy paychecks from a long term Mars mission want to believe.

As a result, "the scientific mind" has been just as guilty as the popular mind in seeing things that aren't there. The illusions hardly ended with Percival Lowell's canal lines, said Murray. He instanced the *Mariner* 6 and 7 flyby missions of 1969, when a scientist on the ground misinterpreted results relayed from an on-board instrument because, said Murray, "the person really wanted to believe that he discovered something that was a real clue to the existence of life on Mars."[39]

It was a false flag, though as with other false positives in the search for life on Mars, it gave a boost, however briefly, to the cause of Martian exploration.

Mariner 9 sent back to Earth over 7,500 pictures of the Martian landscape. It revealed, among other things, that the planet had enormous volcanic features, a promising discovery that deepened the mysteries of a world that has long fascinated those on Earth who gaze skyward. The *Mariner* 9 findings revived the flagging interest of many astronomers in the planet. As Arthur C. Clarke said after the mission, the "depressing image" of Mars from the mid-1960s until *Mariner* 9 had been that of a "cratered, desiccated wilderness… about as far removed from the Lowell–Burroughs fantasy as it was possible to get."[40] It had become — at least to those dreamers who were so desperate to find life or hints thereof on Mars — dull and uninteresting, which is another way of saying that it was getting awfully hard to attract federal funding. Who wants to spend a small mint just to send people up to frolic in dust that may be differently colored from that on the Moon but that is every bit as lifeless?

Bruce Murray was still, by his account, "playing the role of the heavy" after *Mariner 9*, which sent back zero evidence of life now or at any other point in the history of Mars.[41] Murray wondered why this misguided search for life had become so central to the Mars project. Wasn't it enough to learn from robotic or photographic probes of the manifold and, to planetary scientists at least, endlessly fascinating chemical and physical processes by which Mars was formed and developed? Why this need to inject little green men into every discussion of the Red Planet?

It would take a psychologist to speculate on the need, but the result was obvious: only the tantalizing lure of the possibility of finding life on Mars, fed by so many hundreds of books and movies over the decades, could get American taxpayers and their elected representatives to cut open the public purse and authorize the astronomical expenditures necessary for a manned trip to Mars and back. (Or just to Mars, forget coming back, as Werner von Braun had so strangely suggested years earlier.)

Arthur C. Clarke had ended his remarks to the Pasadena audience with the confident prediction, "Whether or not there is life on Mars now, there *will* be by the end of the century."[42]

He was wrong, spectacularly so. The U.S. has not even come close to authorizing a Mars mission, though advocates of such a mission have never let go of the dream, and have found various creative ways to further their goal.

In 1975, the unmanned *Viking 1* and *Viking 2* missions to Mars were space-craft containing both orbiters and landers. The orbiters circled Mars, taking and transmitting photographs of the planet. On the surface, the landers deployed instruments measuring chemical, biological, meteorological, and other scientific data on Mars. Widely regarded as successful probes, the *Vikings* cost taxpayers about $1 billion — or several times less than even the least ambitious manned Mars mission would have cost.

The *Vikings* failed to find evidence of life, whether ancient or contemporary, on Mars, which was a blow to the Men-to-Mars crowd.

Then a series of misbegotten Mars missions — the *Mars Observer*, which went missing in space in August 1993; the *Mars Climate Orbiter*, which crashed into the planet in 1999; *Deep Space 2*, which went silent in 1999 — gave rise to a popular notion that probes of Mars were jinxed, even though other missions — the *2001 Mars Odyssey*, the *Phoenix* of 2008 — performed as expected. Still, the public reaction to the "successful" unmanned voyages has been a big ho-hum. The most interested citizens might spend a few minutes looking at the images on their computers — as long as they don't take too long to download — and then they return to watching *American Idol*. Their imaginations are unfired.

But what really gets the mission to Mars people tingling is the prospect of men and women on Mars, and that is where the oversell, the exaggeration of stakes, comes into play.

Space buffs remain committed to a Red Planet mission, though they are casting about for an effective political strategy. "I can't think of a more exciting prospect than seeing humans walk on Mars in my lifetime," writes *Sky & Telescope* editor Robert Naeye.[43] It is also hard to think of a more expensive prospect in our lifetimes, though that goes unsaid.

All presidents, it seems, have visions of getting credit for Columbus-like expeditions to other planets, even though Ferdinand and Isabella play a rather meager supporting role in the popular conception of the Columbian discovery of the new world. Ronald Reagan asked his NASA administrator, James Beggs, "Why aren't you going all the way to Mars?"[44] Reagan's defense buildup supplied the answer to his question. Even the first George Bush, who famously derided "the vision thing," dreamed Martian dreams, putting his hapless Vice President Dan Quayle in charge of a Space Exploration Initiative (SEI) intended to put Americans back on the Moon in 2000 and on Mars by 2019.

Bush announced this goal on July 20, 1989, the 20th anniversary of *Apollo 11*'s landing on the Moon. Standing in front of the National Air and Space Museum on the Mall in Washington, DC, surrounded by lunar royalty — Neil Armstrong, Buzz Aldrin, and Michael Collins from *Apollo 11* — President Bush declared, "We must commit ourselves anew to a sustained program of manned exploration of the solar system and, yea, the permanent settlement of space." He called for a "manned mission to Mars."[45] Bush did not put a hard and fast date on the Mars landing until a speech in May 1990 at Texas A&M University, when he said, "I believe that before Apollo celebrates the fiftieth anniversary of its landing on the Moon, the American flag should be placed on Mars."[46]

In a fit of purple prose that makes a previous speechwriter, Peggy Noonan, sound like a hardbitten cynic, Bush spoke on the Mall of "10 very special reasons why America must never stop seeking distant frontiers: the 10 courageous astronauts who made the ultimate sacrifice to further the cause of space exploration. They have taken their place in the heavens so that America can take its place in the stars."

Bush asked his "right-hand man, our able Vice President, Dan Quayle," to direct the National Space Council to make concrete recommendations as to how to achieve this and other goals, which included a space station and a return to the Moon.[47] The not terribly able Quayle had a daunting task: a Gallup poll found that popular support for space exploration was "lukewarm at best" and that "the American public favors decreasing NASA's budget." The director of the Annenberg School for Communication at the University of California, Peter Clarke, observed at the time, "People are convinced that space exploration is frivolous, a luxury, a hobby to be indulged only after other societal tasks are well in hand."[48] Clearly a new justification for this perceived frivolity had to be invented. The Berlin Wall was crumbling, the Soviet Union was destined for the rubbish heap of history, and the *Mariner* and *Viking* unmanned missions had established beyond even the doubt of any latter-day Percival Lowells that little green men did not reside on our planetary neighbor.

The "90-Day Report," hastily put together within three months of Bush's directive, proposed a 30-year program that would have been, space-travel advocate Robert Zubrin writes with disgust, "the largest and most costly U.S. government program since World War II." It would have cost a head-spinning $450 billion and sent astronauts in massive dreadnoughts to Mars for a grand total of two weeks. It was a "flags and footprints" plan whose purpose was to provide full employment for aerospace engineers and stick Old Glory in the Martian soil.[49] It was arid, uninspiring, and so outrageously costly that no one ever took it seriously. As vice presidential candidate Senator Lloyd Bentsen (D-TX) famously observed, Dan Quayle was no Jack Kennedy.

The Bush Space Exploration Initiative gave birth to an Exploration Programs Office (ExPO) and a New Initiatives Office (NIO) within NASA; these died bureaucratic deaths in, respectively, December 1992 and March 1994. As his son would later relearn, a president cannot juggle both a foreign war and an interplanetary expedition. Or at least the Congress would not let him.

Bill Clinton seemed less interested in space, though one suspects that if Mars really did contain a bevy of buxom Barsoomites he'd have spared no expense to bring them back to Earth. He did his part to mislead, or try to mislead, Americans into supporting a Martian program, however, when in August 1996 he jumped the gun in proclaiming evidence that life had once existed on Mars.

To be fair to Clinton, NASA had briefed him about a paper that was soon to appear in *Science* titled "Search for Past Life on Mars: Possible Relic Biogenic Activity in Martian Meteorite ALH84001." The paper, written by a team headed by geologist Dave McKay, claimed that fossilized bacteria had been detected within a meteorite found in Antarctica that had originated from Mars. Though the nature of the tiny "worms" or squiggles found within the rock were open to different interpretations, the authors of the *Science* paper made bold to "conclude that they are evidence for primitive life on early Mars."

Clinton dove right into the scientific controversy, despite what may be charitably described as his lack of knowledge of exobiology.

"Today, rock 84001 speaks to us across all those billions of years and millions of miles," said the President, his phraseology reminiscent of Carl Sagan. "It speaks of the possibility of life. If this discovery is confirmed, it will surely be one of the most stunning insights into our universe that science has ever uncovered. Its implications are as far reaching and awe inspiring as can be imagined."[50]

Clinton was trying to associate himself with what he imagined could be a world-historical event. Alas for him, rock 840001 was something of a p.r. dud. Geologists and exobiologists were skeptical of the claims by the *Science* team, and the verdict on rock 84001 has yet to be handed down. But the brief hubbub did raise the profile of Mars and reintroduce the tantalizing possibility that life once existed, if not flourished, on our neighbor. No little green men, no greenbacks was the lesson Martian explorers were learning.

The second President George Bush had Iraqi fish to fry, but even he took time out to propose a vast plan to colonize the solar system.

In the glow of the successful Martian ramblings of the rovers *Spirit* and *Opportunity*, which would send back breathtaking photos of their vacation, in early 2004 President Bush announced what became the Vision for Space Exploration (VSE), its title perhaps a sly (or embarrassed) nod to the president's father's much-noted troubles with "the vision thing."

The VSE, which was pronounced dead even before arrival, would have spent somewhere (no one really knows where) upwards of a trillion dollars sending humans back to the Moon by 2020 and to Mars in the years to follow. The President did not bother to announce that the Moon and Mars possessed weapons of mass destruction, and so the VSE never really showed up on political radar screens dominated by "homeland security" and war.

Mars may be the Roman god of war, but in the 1980s it was seized upon, if briefly, as the instrument of international peace. Liberal internationalists saw it as the heavenly cause toward whose exploration the lamb and the lion, or the United States and the Soviet Union, might usefully cooperate. As Daniel Deudney of the World Policy Institute argued in a 1985 issue of *World Policy Journal*, "global security could be enhanced simply by redirecting

U.S.–Soviet competition from the military arena to civilian activities" – the bigger the better.[51] And few projects would be bigger than a trip to Mars.

Deudney took heart from the *Sputnik* example, which, as he noted, "the Democrats seized upon" as a way to push a program of increased federal spending on education. In recent years, however, the "space movement has increasingly been courted by right-wing extremist, even protofascist groups." The use of such loaded terms does more to discredit Mr. Deudney than his targets, but fear not: he is hopeful of "restoring the traditional link between the space program and progressive social forces."[52] Mars was interesting to such people only insofar as it could be used to further a statist political agenda. It would be, they hoped, explored and colonized by the Soviets and the Americans and one gets the impression that it would be subject to so stultifying and suffocating a set of bureaucratic regulations that if it wasn't already dead it would be thoroughly lifeless after the liberals got through with it.

The "transformative potential of cooperative missions involving astronauts and cosmonauts" might cost hundreds of billions of dollars but at least it would "cultivate the sense of shared human identity and destiny that must eventually replace the outdated provincialism of nationalistic thinking."[53] These transformative missions would be paid for by citizens of two of those outdated nations, though once people had evolved into the possession of Deudney's desired sense of shared human identity no doubt the rest of the peoples of the world would gladly chip in to relieve the Russians and Americans of the outsized due bill.

Yet even so ardent a humans-to-the-stars advocate as the late astronomer and science popularizer Carl Sagan, who had grown up reading Edgar Rice Burroughs and dreaming of Barsoom, was dubious about a manned mission to Mars. "Especially with continuing investment in robotics and machine intelligence, sending humans to Mars can't be justified by science alone," he wrote in *Pale Blue Dot: A Vision of the Human Future in Space* (1994), a sequel, of sorts, to his phenomenal best-seller *Cosmos* (1980).[54]

Sagan, an internationalist and anti-nuclear activist, had once hoped that a joint U.S.–Soviet Mars trip would be an aeronautical means of détente. Whereas a previous generation had relied on the Cold War to generate the fuel that sent rockets into space, Sagan dreamed of U.S.–Soviet cooperation. The *Apollo–Soyuz* Test Project, which was sealed by a "handshake in space," was a start. As a grand sequel he proposed a 1992 mission to Mars to mark the quincentennial of Columbus's trip to (don't dare say "discovery of"!) the New World and the 75th anniversary of the Russian Revolution that brought Lenin and the Communists to power. It is likely that a substantial number of

Russians would have looked frostily upon such a commemoration, but in any event, Sagan's idea never achieved liftoff.

As the Cold War ended, Carl Sagan wrote that the absence of an enemy "made some of the American aerospace industry and some key members of Congress profoundly uneasy. Without international competition, can we motivate such ambitious efforts?"[55] For ideological reasons, Sagan never dropped his insistence upon an international mission to Mars. Human brotherhood, he believed, would be furthered by sending a handful of scientists from the leading industrial nations on a multiyear space trip.

If political liberals such as Sagan see Mars, named after the god of war, as the cementer of peace on Earth, Robert Zubrin, who has friendly relations with the frenetic Republican idea-machine Newt Gingrich, views Mars as our last chance to keep the candle of vitality from blowing out. "Without a frontier from which to breathe new life, the spirit that gave rise to the progressive humanistic culture that America has represented for the past two centuries is fading," Zubrin mourns. Now that the West is settled, Americans — or at least the four who will make a Mars trip — need "a frontier to grow in," lest "the entire global civilization based upon values of humanism, science, and progress will ultimately die."

"I believe," says Zubrin, "that humanity's new frontier can only be Mars."[56] In other words, only by supporting a manned expedition to Mars can a person reveal that he or she is in favor of the continuation of human civilization. Given such a choice, who could possibly dissent?

With billions fewer workers on Mars, Zubrin reasons that this labor shortage will cause wages to be considerably higher on Mars than on Earth. "Workers on Mars will be paid more and treated better than their counterparts on Earth, and education will be driven to a higher standard than ever seen on the home planet." It will be a Democratic Party utopia! And it will filter back to Earth. "A new standard may be set for a higher form of humanist civilization on Mars, and, viewing it from afar, the citizens of Earth will rightly demand nothing less for themselves."

Zubrin is a creative thinker and a burr in the saddle of the NASA bureaucracy, but in holding up Mars as the goad to liberal social and labor reform he is venturing into a kind of out-of-this-world speculation. He is also acting in the time-tested tradition of government-science seekers of funds: claim that the sky is falling — or in this case, human civilization is coming to an end — and that the only thing that can keep the sky above our heads and civilization humming is to adopt the pet cause of the advocate. He concludes, "If the era of Western humanist society is not to be seen by future historians as some kind of transitory golden age, a brief shining moment in an otherwise endless chronicle of human misery, then a new frontier must be opened. Mars beckons."[57]

Support a manned mission to Mars or be in favor of an endless chronicle of human misery. Hmm. That's some choice. Maybe we can see a third option on the menu, please?

If you can't persuade your fellow citizens to expend untold billions of their tax dollars on your pet government projects, then the next best thing to do is insult them, call them names, imply that they are imbeciles and dullards too stupid to grasp the crystalline logic and sophistication of your position. Arthur C. Clarke, in his book *Interplanetary Flight: An Introduction to Astronautics* (1950), asserted: "The choice, as Wells once said, is the Universe — or nothing. . . . The challenge of the great spaces between our worlds is a stupendous one; but if we fail to meet it, the story of our race will be drawing to its close. Humanity will have turned its back upon the still untrodden heights and will be descending again the long slope that stretches, across a thousand million years of time, down to the shores of the primeval sea."[58]

In other words, if you question the need for astronomically expensive space projects you are an advocate of devolution, of regressing from humankind to some earlier evolutionary stage, back to that period in the dawn of time when our ancestors slithered across rocks or swam in primordial soup. You are not just stupid. You are not even a Cro-Magnon, or a Neanderthal. You are pre-human, pre-ambulatory, pre-cognitive. You are not even worthy to take part in a conversation with the likes of Arthur C. Clarke.

But when the insults fail, go back to scaring the rubes.

The Mission to Mars campaign may be one beneficiary of the global warming scare fanned with such media smarts over the last decade-plus by Albert Gore and the well-organized institutions of establishment environmentalism.

Consider *Dire Predictions: Understanding Global Warming* (2008) by Michael E. Mann and Lee R. Kump. In the view of Mann and Kump, global warming is fact, not theory, and climate change is bearing down on us like a runaway train. Only drastic measures can save the earth. They feature the British scientist James Lovelock, author of the best-seller *The Revenge of Gaia* (2006), who warns that "for now, the evidence coming in from the watchers around the world brings news of an imminent shift in our climate towards one that could easily be described as Hell: so hot, so deadly that only a handful of the teeming billions now alive will survive."[59]

Note the dismissive attitude toward human beings in Lovelock's diction: you, me, Barack Obama, Angelina Jolie — we are all ants in the "teeming millions" who deface Gaia. Lovelock's solution to the problem: terraforming

Mars by "melting the ice-caps, which are rich in frozen CO_2, with nuclear warheads. This would release greenhouse gas into the atmosphere and warm the planet to habitable conditions."[60]

In Lovelock's formulation, man has made such a mess of the Planet Earth that it's time for him to blast Mars with nuclear warheads and create our Eden there. Mars is the alternative to a literal Hell on Earth, says Lovelock. The Mars crowd may not really believe this, but it is another useful arrow to pull from the quiver.

The indefatigable aerospace engineer Robert Zubrin, a nuclear engineering Ph.D. and former Lockheed Martin senior engineer, has emerged as the most visible and persuasive advocate for a Martian mission over the last decade. A central figure in the "Mars Underground" and founder of the Mars Society, Zubrin is at the center of an enthusiastic and dedicated band of aeronautical engineers, astronomers, science-fiction writers, filmmakers, businessmen, and students created in 1998 to "further the goal of the exploration and settle-ment of the Red Planet." Though a tertiary goal is the support of "Mars exploration on a private basis," the Mars Society calls for "ever more aggres-sive government funded Mars exploration programs around the world."[61]

Zubrin laid out *The Case for Mars: The Plan to Settle the Red Planet and Why We Must* (with coauthor Richard Wagner) in 1996. Mars Direct, the forceful tag applied to his proposal, called for a four-person crew to make a two-and-a-half year round trip to Mars, with a surface stay of 500 days — far longer than most competing plans call for. The astronauts — Martians, really, for they would live on the planet for a longer stretch than many Americans live in their homes or apartments — would tool around in "Mars cars" and live off the land, in a manner of speaking. Even the propellant which would return the astronauts to Earth would be derived from Martian resources, as would other consumables. Zubrin confidently predicts, "We can make rocket fuel and oxygen on Mars."[62] Taking inspiration from those Arctic explorers who packed lightly and took their sustenance from the country surrounding them, Zubrin's "dog sled" approach would slash dramatically the cost of a Mars mission. He estimated in the late 1990s that cost to be about $25 billion — a considerable sum, to be sure, but a dust mite com-pared to the $450 billion price tag of the first President Bush's SEI.[63]

The Mars Underground, whatever one might think of the wisdom or jus-tice of spending tens of billions of taxpayer dollars on such a mission, is a fascinating grass-roots movement that sprang up in Boulder, Colorado, in the late 1970s, as a group of imaginative and enterprising graduate students at the University of Colorado gathered together like-minded folk to talk about the possibilities of landing human beings on Mars and then "terraforming" the planet – that is, making it more Earth-like in order to be conducive to colonization. Andrew Chaikin has told the story with great panache in *A Passion for Mars* (2008) of how these grad students and "Closet Martians" from NASA and the aerospace industry met and brainstormed and dreamed toward "one major goal: landing of the human species on, and eventual occupancy of, the Red Planet Mars," in the words of an early manifesto.[64] One of the Mars Underground's first NASA members was Thomas Paine, who at least is no sunshine patriot of the interplanetary manned travel crowd.

Mars Direct was formulated about a decade later. It was seeded in the Mars Underground and came to fruition as a response to the crashing failure of Bush's Space Exploration Initiative, whose lumberingly expensive Mars mission seemed to kill the idea of a visit to the Red Planet for at least a generation.

The Mars Underground and Zubrin's Mars Direct envision a Mars that is colonized and terraformed into a kind of Earth II. This vision has even won converts at NASA, though some within the agency snipe at it because it makes no use of the Space Station upon which so many aerospace contracts in the districts of so many influential congresspersons depend.

Meeting in College Park, Maryland, just outside Washington, DC, in the summer of 2009 the Mars Society organized the Second Great Mars Blitz, which launched upwards of 100 conferees toward Capitol Hill, where they lobbied for human exploration of Mars. That a movement that began with underground enthusiasm has bubbled up, or curdled, into a lobby is a sad commentary on idealism in modern American politics.

For those who share the bedazzlement over Mars but would rather not authorize the federal government to spend hundreds of billions, perhaps trillions, on a manned Mars mission, there is an alternative. Pioneer nanotechnologist Eric Drexler, holder of the first Ph.D. in Molecular Nanotechnology ever granted by the Massachusetts Institute of Technology, laid it out in the October 1984 issue of *L5 News*. The L5 Society had been founded in 1975 to promote the colonization of space. Though it later merged with the Werner von Braun-founded National Space Institute to form the National Space Society, for about a dozen years L5 acted as a creative force in the avant-garde of space-travel theorists. Its members were often harsh critics of NASA.

Dr. Drexler had little use for gaseous Saganesque paeans to the ineffable value of spacecraft floating out there amidst the billions and billions of stars. "To open space to settlement, we must use space for practical purposes," he declared. "What could be more obvious?" Mining and agriculture had opened the American West to settlement, and Drexler envisioned the same process at work in the solar system and beyond. He saw potential in the metals of the Moon, which so many others had dismissed as a barren and lifeless rock, and he positively thrilled to the riches of asteroids, with their treasures of "oxygen, rock, water, hydrocarbons, steel, nickel, cobalt, and precious metals."

But "Mars is not even in the running," he said. It has resources, to be sure, but they pale in comparison with those offered by the asteroids, many of which are considerably more accessible. Asteroid mining might well become attractive to private industry, which would assume the risks involved. In time, those who do the mining would undoubtedly set up settlements on those asteroids offering the most plenteous resources. In this way would other worlds be colonized. The planets, we may expect, would follow — in due time, and in the more or less natural course of events, instead of in a fiscally (and possibly physically) reckless crash program.

After rehearsing his arguments that the Martian atmosphere would pose nearly insuperable obstacles to human settlement in the foreseeable future, Drexler takes up — and takes down — the ethereal "We must go because we must go" colonizers. Without any economic incentive to settle Mars, housing on the planet "must be built using tax money and maintained using still more tax money." Such a colony would be — a colony. It would lack even the façade of independence. It would be utterly, hopelessly dependent upon Earth subsidies.

Responding to a pro-Mars reader who remonstrates that "Some of us dream of living on other planets," the coolly logical Drexler says, "No doubt this is so, but it makes sense to adapt dreams to opportunities. Some of us may dream of receiving federally-funded annuities of ten million dollars a year, but it makes no sense to ask for it." The goal, says Drexler, should be "to open space as a practical frontier — not to plead for high-cost public housing on other planets."

Private enterprise, not taxpayer-fed fantasies of national glory, should drive space exploration, argues Drexler. We don't need "grand stunts" that lead down gold-paved paths to glory. The cynical strategy of the Mars crowd he derides as "whip[ping] up the public and siphon[ing] off tax money." Their desire is for a "political stunt, with some hope of setting up an open-ended charity." There is nothing of liberty or freedom in such a scheme; it is little more than a flashy scam, a hornswaggling of the public. Drexler characterizes the mission-to-Mars argument as "We're for space, and the best thing we can think of doing is to land a guy in a space suit on yet another cratered ball of rock, and maybe hang out there for a long time… isn't that great? Just cough up the bucks!"[65]

No thanks. The public has long since lost interest in the space program. When Neil Armstrong bounded from *Apollo 11*, "94 percent of homes with televisions were tuned to the space event," as Leonard David of Space Data Resources & Information noted.[66] Yet subsequent missions to the Moon were regarded as so ho-hum that NASA had to beg for television coverage. Tom Wolfe, in *The Right Stuff*, made famous the lesson in lobbying that the *Mercury* astronauts taught NASA and the military–space–industrial complex: No Buck Rogers, no bucks. If Americans had to pay for a space program that most thought a luxury of marginal interest, at least they wanted dashing astronauts who performed acts of derring-do.

The Mars mission folks understand this, though the cost of such a mission, unless it is conducted along Zubrin lines, is so high that Buck Rogers is eclipsed by the bucks.

Absent Buck Rogers, even the relatively less expensive robotic missions struggle for funding — although the NASA bureaucracy, born in the *Sputnik* panic, which recently turned 50 years of age, has achieved a level of permanence that guarantees a flow of money, even in the worst of times. The current Great Red Hope is the *Mars Science Laboratory*, or the "mega-rover," which at 1,875 pounds is heavier than the combined weights of the two most recent Mars rovers, *Spirit* and *Opportunity*. (These last two sound as if they were named by Newt Gingrich, or at least NASA publicity people conscious of the need to appeal to Republican congressmen with GOP buzzwords.)

Mars Science Laboratory, with its payload of ten instruments, has seen its launch delayed from 2007 until at least 2011. Cost overruns have boosted its bill to $2.2 billion and counting, or almost 30 percent above what it was projected to cost. It is not the magic key to a human mission to Mars.[67]

Even Andrew Chaikin, who has written with great sympathy of those who have devoted their lives to the study of and possible journeys to Mars, concedes that "The visions of space artists and Hollywood filmmakers aside, landing humans on Mars is simply not doable — at least, not yet… von Braun's statement in 1954 that it would be a hundred years or more before humans were ready to go to Mars seems right on target, maybe even optimistic."[68]

The only choice the would-be Martians have is to create a panic. They haven't figured out how to do that yet — though they could take a lesson from our next subject. For if the case of *Sputnik* is an example of how advocates of federal subsidy got what they wanted, and the case of a manned mission to Mars is an example of a failed — or at least uncompleted — quest for federal funds, the case of the killer asteroid falls — with an out-of-this-world *thud* — somewhere in between.

Notes

1. Stephen Baxter, "How NASA Lost the Case for Mars in 1969," *Spaceflight*, Vol. 38 (June 1996), p. 191.
2. Quoted in Richard W. Orloff and David Michael Harland, *Apollo: The Definitive Sourcebook* (New York: Springer-Praxis, 2006), p. 12.
3. Baxter, "How NASA Lost the Case for Mars in 1969," p. 191.
4. Quoted in Sagan, *Pale Blue Dot*, p. 189.
5. Annie Platoff, "Eyes on the Red Planet: Human Mars Mission Planning 1952 1970," NASA CR-2001 208928, July 2001, p. 2.
6. Ibid., p. 4.
7. Andrew Chaikin, *A Passion for Mars: Intrepid Explorers of the Red Planet* (New York: Abrams, 2008), p. 43.
8. Werner von Braun, "The Next 20 Years of Interplanetary Exploration," *Astronautics & Aeronautics* (November 1965), p. 24.
9. Ibid., p. 33.
10. Ibid., p. 34.
11. Robert Zubrin with Richard Wagner, *The Case for Mars: The Plan to Settle the Red Planet and Why We Must* (New York: Free Press, 1996), p. 47.
12. Platoff, "Eyes on the Red Planet: Human Mars Mission Planning, 1952–1970," pp. 14–15.
13. Ibid., p. 18.
14. Ibid., p. 25.
15. Ibid., pp. 30, 32.
16. Chaikin, *A Passion for Mars: Intrepid Explorers of the Red Planet*, p. 53.
17. Baxter, "How NASA Lost the Case for Mars in 1969," p. 192.
18. Platoff, "Eyes on the Red Planet: Human Mars Mission Planning, 1952–1970," p. 46.
19. Linda R. Cohen and Roger G. Noll, *The Technology Pork Barrel* (Washington, DC: Brookings Institution, 1991), p. 180.
20. Baxter, "How NASA Lost the Case for Mars in 1969," p. 191.
21. Chaikin, *A Passion for Mars: Intrepid Explorers of the Red Planet*, p. 37.
22. Ibid.
23. Baxter, "How NASA Lost the Case for Mars in 1969," p. 192.
24. Thomas O'Toole, "Controversy Expected on Mars Landing Goal," *Washington Post*, August 10, 1969.
25. Quoted in Chaikin, *A Passion for Mars: Intrepid Explorers of the Red Planet*, p. 40.
26. Platoff, "Eyes on the Red Planet: Human Mars Mission Planning, 1952–1970," pp. 48–49.

27. Baxter, "How NASA Lost the Case for Mars in 1969," p. 192–93.
28. Platoff, "Eyes on the Red Planet: Human Mars Mission Planning, 1952–1970," p. 49.
29. Baxter, "How NASA Lost the Case for Mars in 1969," p. 191.
30. "Poll Finds Public Cool to Mars Trip," *New York Times*, August 8, 1969; Baxter, "How NASA Lost the Case for Mars in 1969," p. 193.
31. Chaikin, *A Passion for Mars: Intrepid Explorers of the Red Planet*, p. 49.
32. Sylvia K. Kraemer, "Opinion Polls and the US Civil Space Program," *Journal of the British Interplanetary Society*, Vol. 46 (November 1993), p. 445.
33. *Congressional Record*, August 13, 1969, p. 23805.
34. Kraemer, "Opinion Polls and the US Civil Space Program," pp. 444, 446.
35. Platoff, "Eyes on the Red Planet: Human Mars Mission Planning, 1952–1970," p. 50.
36. Baxter, "How NASA Lost the Case for Mars in 1969," pp. 192, 194.
37. Zubrin with Wagner, *The Case for Mars: The Plan to Settle the Red Planet and Why We Must*, p. 27.
38. Ray Bradbury, Arthur C. Clarke, Bruce Murray, Carl Sagan, and Walter Sullivan, *Mars and the Mind of Man* (New York: Harper & Row, 1973), p. 13.
39. Ibid., pp. 22–23.
40. Ibid., p. 83.
41. Ibid., p. 47.
42. Ibid., p. 28.
43. Robert Naeye, "Humans to the Moon?" *Sky & Telescope* (March 2009), p. 8.
44. Quoted in Chaikin, *A Passion for Mars: Intrepid Explorers of the Red Planet*, p. 123.
45. George Bush, "Remarks on the 20th Anniversary of the *Apollo 11* Moon Landing," July 20, 1989, The American Presidency Project, http://www.presidency.ucsb.edu/ws.
46. Quoted in Leonard David, "Mars: the Media…the Masses…and the Message," in *Strategies for Mars: A Guide to Human Exploration*, edited by Carol R. Stoker and Carter Emmart (San Diego: American Astronautical Society, 1996), p. 41.
47. Bush, "Remarks on the 20th Anniversary of the *Apollo 11* Moon Landing."
48. David, "Mars: the Media…the Masses…and the Message," p. 42.
49. Zubrin with Wagner, *The Case for Mars: The Plan to Settle the Red Planet and Why We Must*, pp. 46–47.
50. Chaikin, *A Passion for Mars: Intrepid Explorers of the Red Planet*, pp. 157–58.
51. Daniel Deudney, "Forging Missiles into Spaceships," *World Policy Journal*, Vol. 2, No. 2 (Spring 1985), p. 276.
52. Ibid., pp. 278–79.
53. Ibid., pp. 292, 303.
54. Sagan, *Pale Blue Dot*, pp. 250–51.
55. Ibid., p. 253.
56. Zubrin with Wagner, *The Case for Mars: The Plan to Settle the Red Planet and Why We Must*, p. 297.
57. Ibid., pp. 302, 304.
58. Arthur C. Clarke, Foreword, *The Case for Mars*, p. xi.

59. Michael E. Mann and Lee R. Kump, *Dire Predictions: Understanding Global Warming* (New York: DK Publishing, 2008), p. 120.

60. Ibid., p. 121.

61. http://www.marssociety.org.

62. Zubrin with Wagner, *The Case for Mars: The Plan to Settle the Red Planet and Why We Must*, p. 156.

63. Ibid., p. xix.

64. Chaikin, *A Passion for Mars: Intrepid Explorers of the Red Planet*, p. 136.

65. Eric Drexler, "The Case Against Mars," *L5 News*, October 1984, www.nss.org, *passim*. Carl Sagan disposed easily of the ridiculous spin-off argument: that it was worth going to the Moon because we got Tang out of it. "Spend $80 billion (in contemporary money) to send Apollo astronauts to the Moon, and we'll throw in a free stickless frying pan. Plainly, if we're after frying pans, we can invest the money directly and save almost all of the $80 billion." Sagan, *Pale Blue Dot*, p. 272.

66. David, "Mars: the Media…the Masses…and the Message," p. 44.

67. Joel Achenbach, "New Mars Rover Too Costly, Critics Say," *Washington Post*, February 15, 2009. See also "Martian Mega-Rover Delayed," *Sky & Telescope* (March 2009), p. 17.

68. Chaikin, *A Passion for Mars: Intrepid Explorers of the Red Planet*, p. 269.

Chapter 6

The Chicken Littles of Big Science; or, Here Come the Killer Asteroids!

Carl Sagan admitted, "On first hearing about the Asteroid Hazard, many people think of it as a kind of Chicken Little fable."[1]

That they do, but Chicken Little, he of "the sky is falling!" fame, never got the Hollywood treatment in the way that the "asteroid hazard" has. The falling sky sells a planet's worth of DVDs.

"If you want to send a message, call Western Union," famously said either or both Sam Goldwyn and Jack Warner, the movie moguls who share credit for this quotable quote. And while this may be good advice at the box office, in the late 1990s Hollywood did a send a message about Big Science to the popcorn-munchers in the dark: Be afraid. Be very afraid. The killer asteroids are coming. And as a corollary: Fund asteroid hunters. Your life — and the future of the entire planet — may depend on it.

This message, however distorted the details were once they hit the screen, was welcomed by one of the newest fields in end-of-the-world-prevention: astronomers who keep watch on comets, asteroids, and other bodies that may, at some time over the next 100 million or so years, deliver a cosmic concussion (or worse) to the Earth.

First, the films, which came out one after the other. *Deep Impact* (1998) was director Mimi Leder's big-budget scareathon about a killer-comet whose approach to earth can only be stopped by a spaceship crew of young models, with the assistance of the grizzled veteran Robert Duvall. President of the United States Morgan Freeman declares martial law. Government loves emergencies, after all, and as Obama chief of staff Rahm Emanuel said in that endlessly repeatable remark, you should never waste a good crisis. The coasts

J.T. Bennett, *The Doomsday Lobby: Hype and Panic from Sputniks, Martians, and Marauding Meteors*, DOI 10.1007/978-1-4419-6685-8_6, © Springer Science+Business Media, LLC 2010 are wiped out by tsunamis when a fragment of the comet hits the earth. The courageous models in space, however, destroy the major part of the comet, and in a final scene we learn that the federal government has survived — and that, really, is the main thing, isn't it? No matter how many millions of workaday Americans get swamped, as long as the Department of Commerce and the Federal Aviation Administration are still humming along, civilization is safe.

A rival doomsday film, Michael Bay's *Armageddon*, also released in 1998, was hokey and entertaining and at a cost of $140 million, it could have covered the cost of all the near-earth object-tracking programs then extant. The film begins with narrator Charlton Heston rehearsing the tale of the asteroid that caused the Cretaceous–Tertiary extinction and concluding, "It happened before. It will happen again. It's just a question of when."

"When," in *Armageddon*, is 18 days from the discovery that a rogue asteroid the size of Texas is headed for Earth. Its arrival will mark "the end of mankind."

"We didn't see this thing coming?" an agitated President asks the head of NASA, played by Billy Bob Thornton.

Thorton's response is music to real-life asteroid hunters' ears. "Our object collision budget is a million dollars," he says. "That allows us to track about 3 percent of the sky, and it's a big-ass sky."

Message received. Boost that budget! Or else Earth and everything on it will be pulverized someday by a big-ass asteroid.

An oil-rig crew, led by Bruce Willis, saves the day by drilling 800 feet into the asteroid and delivering into its belly a nuclear device. They are aided by a vessel that was designed for the "Mars project" — a nice example of aeronautics/astronomy fund-seeking symbiosis!

Armageddon was something of a hit, and it is great fun to watch, but it would pale beside the real thing. Then again, there hasn't been a "real thing" for about 65 million years, so we and the media sensationalists have had to make do with overblown stories of "near misses."

For instance, on March 11, 1998, Brian Marsden, director of the partially NASA-funded Minor Planet Center of the International Astronomical Union in Cambridge, Massachusetts, sent out a circular to amateur and professional astronomers warning that on October 26, 2028, at 1:30 Eastern Standard Time, asteroid 1997 XF11, which was nearly one mile across and had been discovered the previous December by the Spacewatch program, would come dangerously close to the Earth: within 15,000 miles. That it would approach closer to the Earth than even the Moon was "virtually certain," said Marsden.[2]

If it struck our planet, as science writer Martin Gardner wrote, "the devastation would be too awful to contemplate."[3]

Marsden published this circular "[w]ithout checking with any outside scientists," complained astronomer Clark R. Chapman.[4] Although the circular was directed mainly to astronomers, non-astronomers were also on the reception list. These included news-gathering organizations. The potential demolition of Earth is, these organizations gathered, news. And so they reported it.

"[F]or 24 hours," wrote editor Bonnie Bilyeu Gordon of *Astronomy* magazine, "the whole world feared."[5] Were earthlings to go the way of the dinosaurs in October 2028? Did a 30-year warning time give us enough time to plot out a defense or even plan an escape? Or were we to spend the next three decades in a state of more or less permanent panic, biding our time and just waiting for the big one to smash us to bits?

Two days later, Marsden had to say "Oops." Based on a quick check of old photos by Eleanor Helin and others at the Jet Propulsion Laboratory, Marsden recalculated the figure of 15,000 miles as, instead, 600,000 miles. Sorry. Go back to your lives.

Brian Marsden had a henhouse full of egg on his face, and he took his share of criticism from the astronomical community for what many observers considered headline-chasing sensationalism. In fact, just 5 hours after Marsden's circular had gone out to a trembling world, Don Yeomans of the Jet Propulsion Laboratory sent an email to colleagues assessing the chances of XF11 hitting the Earth in 2028 at "zero, folks." Daniel Schorr of National Public Radio asked, "Couldn't they have waited a few days before scaring us half to death?"[6]

But the man who panicked the world looked at the bright side: "The good news in all this," he said in 1998, "is that NASA just announced it's doubling the amount of money it spends on asteroid searches — from $1.5 million a year to $3 million. Perhaps that decision is the result of the public interest in asteroid 1997 XF11, who knows?"[7]

In other words, NASA may well have doubled its spending on a government program due to the panic engendered by a woefully errant report, a classic case of Bad Science. This ought to be cause for dismay. Taxpayers' money is being wasted because of a gross error and the attendant publicity. But instead, this is "good news" because — well, because taxpayers' money is being spent due to a gross error and its attendant publicity. (In fairness to Marsden, MIT astronomer Richard Binzel had pegged XF11 as a rather threatening 3.5 on his 0–10 Torino scale — the Richter Scale of asteroid impacts.)

You can't buy publicity of the sort that XFII gave the cause of doomsday. "Kiss Your Asteroid Goodbye!" declared the *New York Post* after the fact. Yes, there was a giggle factor, but the threat had been raised, and even if it was dismissed, the idea had been planted in millions of heads. The then-head of NASA, Daniel Goldin, found that after the XFII scare, killer asteroids became the "second most common subject of public communication" – second only to the alleged "Face on Mars."

Clark R. Chapman has written that several observers within what is awkwardly known as the "hazards community" have argued that "Marsden should be praised, rather than criticized, for the XFII announcement because the higher public visibility may yield increased funding of the observers' programs."[8] That, in a nutshell, is the problem that skeptics have with the whole doomsday asteroid/comet industry. An exceedingly remote threat is periodically blown up in order to win taxpayer funding of favored projects.

Philip Plait, who debunks pseudoscience from his website Bad Astronomy, notes that "Anytime an asteroid is predicted to pass by the Earth they envision an apocalyptic scenario, fueled by reporters who play up the danger without mentioning the odds of our getting hit are less than the chance of winning the typical lottery."

Even Plait can't resist the imagery of a violent universe, though, as he writes, "The Earth sits in a cosmic shooting gallery, and the Universe has us dead in its crosshairs."[9]

In one respect, the alarmists have won. "There used to be high giggle factor among members" of Congress when the subject of killer comets and asteroids was raised, Richard Obermann, staff director of the House Subcommittee on Space and Aeronautics, said in 2008. "But it's now a very respectable area of investigation."[10] After all, that's how the dinosaurs met their maker, or so many physicists and geologists believe.

From the fossil record, scientists believe that over the last 570 million years there have been five distinct episodes of mass extinction, the most recent occurring 65 million years ago, as the Cretaceous period gave way to the Tertiary period. (The previous four episodes were the Ordovician, Devonian, Permian, and Triassic.) None is believed to have killed off every form of life on Earth, though the Permian–Triassic event seems to have led to the disappearance of as high as 95% of all species. During the K–T Extinction, as it is known (K being the first letter of *Kreidezeit*, the German word for Cretaceous), "the marine reptiles, the flying reptiles, and both orders of dinosaurs died out," in the words of the most noted modern explainers of this fact. "Dramatic extinctions occurred among the microscopic floating animals and plants" as well, though snakes, land plants, mammals, crocodiles, and various invertebrates survived, especially those smaller mammals that fed on "insects and decaying vegetation." In all, "about half of the genera living at that time perished during the extinction event."[11]

What killed the dinosaurs? is one of those seemingly eternal scientific questions that engage the specialist, the generalist, the intelligent reader, and even children. The answers over the years have ranged from a drop in sea level to massive volcanic eruptions and consequent blockage of sunlight, but for the last three decades the most commonly accepted, if still disputed, hypothesis is that an asteroid done it.

In a sense, the current vogue for doomsday scenarios was born of a seminal paper by Luis W. Alvarez, Walter Alvarez, Frank Asaro, and Helen V. Michel in the June 6, 1980, number of *Science*. Titled "Extraterrestrial Cause for the Cretaceous–Tertiary Extinction," the article amounted to a paradigm shifter. Luis Alvarez was a Nobel Prize-winning physicist, his son was a geologist, and the other two authors were chemists. The paper adduced physical evidence for an extinction event that was quite literally out of this world.

Alvarez et al. discovered that iridium, an element common in meteorites but rare on earth, was found at uncommonly high levels in rock strata of 65 million years ago. They hypothesized that the dinosaur-killing body's impact threw rock upwards, and since what goes up must come down (except a portion of the ejecta that was hurled into space) it rained down upon the dinosaurs, killing all but the burrowed smaller creatures, many of which starved to death since most plant life was burned off the earth. Upon impact, the pulverized rock – "60 times the object's mass"[12] – would have been blasted into the stratosphere, where it blotted the sun and blocked photosynthesis. The forests burned, and the soot, too, blocked the healthy rays of our local star. The K–T event created so thunderous a sonic boom that "Any creatures within a thousand miles that survived the initial impact were quite deaf once the thunderclap reached them."[13]

The authors describe it this way: "our hypothesis suggests that an asteroid struck the earth, formed an impact crater, and some of the dust-sized material ejected from the crater reached the stratosphere and was spread around the globe. This dust effectively prevented sunlight from reaching the surface for a period of several years, until the dust settled to earth. Loss of sunlight suppressed photosynthesis, and as a result most food chains collapsed and the extinctions resulted."[14] (The likeliest crater caused by this hypothesized K–T extraterrestrial killer is the Chicxulub off the Yucatan Peninsula, which has been dated at 65 million years old. There are more than 100 terrestrial impact craters, or craters dug out on Earth by impacts. The most famous, and the first recognized as such, is Meteor, aka Barringer, Crater in Arizona.)

It should be noted that the Alvarez et al. hypothesis was not universally accepted. As Peter M. Sheehan and Dale A. Russell wrote in their paper "Faunal Change Following the Cretaceous–Tertiary Impact: Using Paleontological Data to Assess the Hazards of Impacts," published in *Hazards Due to Comets & Asteroids* (1994), edited by Tom Gehrels, "many paleontologists resist accepting a cause and effect relationship" between the iridium evidence, the Chicxulub crater, and the mass extinction of 65 million years ago.[15] For instance, Dennis V. Kent of the Lamont–Doherty Geological Observatory of Columbia University, writing in *Science*, disputed that a high concentration of iridium is necessarily "associated with an extraordinary extraterrestrial event" and that, moreover, "a large asteroid... is not likely to have had the dire consequences to life on the earth that they propose."[16]

Briefly, Kent argues that the Alvarez team mistakenly chose the 1883 Krakatoa eruption as the standard from it extrapolated the effects of stratospheric material upon sunlight. Yet Krakatoa was too small a volcanic eruption from which to draw any such conclusions; better, says Kent, is the Toba caldera in Sumatra, remnant of an enormous eruption 75,000 years ago. (A caldera is the imprint left upon the earth from a volcanic eruption.) The volume of the Toba caldera is 400 times as great as that of Krakatoa – considerably closer to the effect that an asteroid impact might have. Yet the sunlight "attenuation factor [for Toba] is not nearly as large as the one postulated by Alvarez *et al.* for the asteroid impact." Indeed, the Toba eruption is not associated with any mass extinctions, leading Kent to believe that "the cause of the massive extinctions is not closely related to a drastic reduction in sunlight alone."[17]

Reporting in *Science*, Richard A. Kerr wrote that "Many geologists, paleontologists, astronomers, and statisticians... find the geological evidence merely suggestive or even nonexistent and the supposed underlying mechanisms improbable at best." Even the iridium anomalies have been challenged: Bruce Corliss of the Woods Hole Oceanographic Institute argues that the major extinctions associated with the K–T event were not immediate and catastrophic but "gradual and apparently linked to progressive climate change."[18]

Others argue that a massive volcanic event predating the Alvarezian killer asteroid created an overwhelming greenhouse effect and set the dinosaurs up for the knockout punch. A considerable number of scientists believe that gradually changing sea levels were the primary cause of the K–T Extinction. If either of these hypotheses is true – and a substantial number of geologists hold these positions — then the "killer asteroid" is getting credit that it does not deserve.

Even if the K–T Extinction was the work of a rock from space, the Alvarez team credits a "probable interval of 100 million years between collisions with 10-km-diameter objects."[19] The next rendezvous with annihilation won't be overdue for about 40 million years. We have time.

We also have fictional accounts of modern space intruders.

As Martin Gardner pointed out, the subject of a collision between earth and celestial traveler has given us several works of art — of varying degrees of merit.

Edgar Allan Poe's short story "The Conversation of Eiros and Charmion" concerned the "fiery destruction" of the earth by a comet. "We had long regarded the wanderers as vapoury creations of inconceivable tenuity, and as altogether incapable of doing injury to our substantial globe, even in the event of contact," says the narrator.[20] How wrong they were. For the planet was consumed by "A combustion irresistible, all-devouring, omni-prevalent, immediate; — the entire fulfilment, in all their minute and terrible details, of the fiery and horror-inspiring denunciations of the prophecies of the Holy Book."[21]

In H.G. Wells's lesser-known novel *In the Days of the Comet* (1906), human-kind is jolted out of its greed, heartlessness, and selfishness by the close passing of a comet. The green vapor of its tail envelops the earth for 3 hours, putting everyone everywhere to sleep, and when the earthlings awake they are magically altered to behave just as the socialist Wells would wish them to behave. They erect a "World State" and discard the quaint idea of "owner-ship" and people live docilely ever after.[22]

The best-selling novel *When Worlds Collide* (1933) by Philip Wylie and Edwin Balmer was made into a 1951 movie in which a small band of pio-neers escapes Earth before its collision with a breakaway planet. Fortunately for posterity, the survivors include the beautiful Barbara Rush.

These fictions scant asteroids, which are much less lovely than comets. Asteroids, which "resemble misshapen potatoes"[23] and are usually more pockmarked than an adolescent's face, are small bodies composed of rock and metal which orbit the sun. The total mass of all known asteroids is less than that of our moon.

Ceres, the first asteroid discovered, floated into human view on January 1, 1801, a happy new-year gift for its discoverer, Giuseppe Piazzi. Ceres, a body 950 meters in diameter, was in orbit between Mars and Jupiter, and thus posed no conceivable threat to Earth. It was almost a century later that the first "near-earth asteroid," or NEA, was discovered: the wishfully named Eros, in 1898. Near-earth asteroids are those whose orbits bring them, at perihelion (closest approach to the Sun) within 1.3 AU. (An AU is the distance from the Earth to the Sun: 93 million miles, or 150 million kilometers.) Thus they either cross the orbit of the Earth or nearly cross it. As far as is known, not a single one poses any conceivable danger to Earth.

Asteroids should not be mistaken for meteors, those falling stars which burn up harmlessly in our atmosphere. (Those whose surviving fragments reach the ground are called meteorites.) Every day about 100 tons of interplanetary dust, the remnants of comets and meteors, fall to Earth — in dust and speck, and to no one's discomfort.

Well, maybe to one woman's discomfort. On November 30, 1954, 31-year-old Ann Hodges of Sylacauga, Alabama, was hit by an 8.5 pound meteorite as she catnapped on her couch. The space invader came through the roof — hurtling at an estimated several hundred miles per hour — hit a wooden radio cabinet, and struck the unfortunate Mrs. Hodges. Or maybe she wasn't so unfortunate: she was bruised but not too much worse for the wear. Mrs. Hodges was acutely uncomfortable with her new fame as the only known example of a human being who was struck by a meteorite.

The H.G. Wellsian doomsday object was a comet, one of those dirty iceballs of the sort rotten kids throw in late February. Comets, long regarded as heralds of doom, are bodies of ice and dust particles trailed by often spectacular tails. What glorious sights they make in the night sky.

The Oort Cloud, theorized by Dutch astronomer Jan Oort in 1950, is the birthplace of comets, whose spectacular appearance belies a rather common admixture of ice and dirt. The number of comets in the Oort Cloud is thought to be 10 to the 11th power. These would be "long period comets," or those that do not make full orbits of the Sun in less than 200 years. In other words, we don't really know they are coming until we pick them up, for the first time, in a telescopic field. In the case of the chimerical "killer comet," this could be only weeks before it plunges Earthward.

One assertion of the doomsday scenarists is that in our travels through the universe, the solar system may pass close enough to extra-solar system objects — say, stars — that these will exert a gravitational sway over Oort Cloud objects and push them into potentially dangerous new orbits. This is pure speculation. Many astronomers doubt that a Planet X or celestial intruder could shake loose the comet cloud, sending killer-comets toward the Earth. Another conjecture, this one published in *Nature* by Marc Davis and Richard A. Muller of the University of California at Berkeley and Piet Hut of the Institute for Advanced Study at Princeton, has it that periodic comet swarms are triggered by a journey through the Oort Cloud by an unseen companion to the Sun. The authors are appropriately modest: "The major difficulty with our model," they admit, "is the apparent absence of an obvious companion to the Sun," though they gamely suggest that "If and when the companion is found, we suggest it be named Nemesis, after the Greek goddess who relentlessly persecutes the excessively rich, proud, and powerful. We worry that if the companion is not found, this paper will be our nemesis."[24]

Would that all writers of scientific papers had such a sense of humor!

The phantom star Nemesis has not, of this writing, been found. But we have plenty of time. After all, David, Hut, and Muller say that given their calculations, a rogue comet unleashed by Nemesis will "present no danger to the Earth until approximately AD 15,000,000" – or 15 million years from now.[25]

Others in the rogues' gallery of fancied Earth-killers include those small bodies out beyond the eighth planet, trans-Neptunian objects, or TNOs, which sojourn frozenly in what is called the Kuiper Belt. Duncan Steel, among the most ardent Cassandras of doomsday, warns, "It seems inevitable that from time to time a massive TNO must enter the planetary region, possibly to be flung toward the terrestrial planets." We haven't seen one yet. But they're out there. And "Nowhere is safe."[26]

Least safe of all is Jupiter, the fifth planet from the Sun, as we have seen recently.

The earthling exodus of *When Worlds Collide* did not make haste for Jupiter, fortunately, for that giant planet suffered its own catastrophe with the impact of Comet Shoemaker–Levy 9 in 1994.

The comet, or rather the several fragmentary comets that made up the comet, was discovered in March of the previous year by Eugene and Carolyn Shoemaker and David Levy at the Palomar Observatory. It was a most unusual comet, orbiting as it did a planet: Jupiter. Astronomers theorized that the comet had been captured by Jupiter's orbit within the past quarter-century. Its orbit was bringing it closer and closer to its planetary dance partner, until its fragments struck Jupiter over the course of a week (July 16–22, 1994). The serial impacts were spectacular, setting off seismic waves, extraordinary atmospheric and chemical disturbances, and leaving dark scars thousands of miles across on the planet. They also set the stage for impact lobbyists to assert that we Earthlings "live in a cosmic shooting gallery and that being hit by a 'big one' is simply a matter of time."[27]

As Carl Sagan notes, the same week that Comet Shoemaker–Levy 9 crashed into (and was absorbed by) Jupiter, the House Science and Space Committee hurriedly prepared legislation directing NASA and affiliated agencies to find and catalogue "comets and asteroids that are greater than 1-km in diameter" and are either in or heading toward the general neighborhood of Earth.[28] Shoemaker–Levy 9 had its destructive side, to be sure, but it was constructive, too, at least for those seeking federal funds for comet and asteroid searches — both the responsible scientists who believe in watchfulness and vigilance and the scaremongering Chicken Littles who would profit from panic. Comet Shoemaker–Levy 9 was a most vivid demonstration that comets can and do strike — though not destroy – planets.

When in July 2009 Jupiter was visited by yet another comet, this one about a kilometer or so in size, Yahoo!News took the bait and observed, in the lede of a story by Charles Q. Choi, that "our solar system is a shooting gallery that sometimes blasts Earth." (The facts that Jupiter has a mass 318 times that of Earth and is adjacent to the asteroid belt went undisclosed in the piece.)

"Shooting gallery" seems to be the metaphor of choice, and Mr. Choi did not disappoint. He quoted Donald Yeomans of NASA's Near-Earth Object project at Cal Tech's Jet Propulsion Laboratory (JPL) speculating that "If an object of about the same size that just hit Jupiter also hit Earth… it would have been fairly catastrophic."[29]

Yes, and if Earth had a silicon-based atmosphere you wouldn't be reading this. The point is that it didn't strike, nor has anything remotely near that size struck Earth in recorded history.

The closest thing to an impact even distantly related to the "catastrophic" occurred just over a century ago. In June 1908, in an event that is central (because seemingly unique in modern times) to the killer asteroid/comet lobby, the so-called Tunguska asteroid, 70 yards (60 meters) in length, exploded about 8 kilometers above the ground in remote Siberia. Its explosion unleashed 20 or more megatons of energy and "flattened about 2,000 square kilometers of forest."[30] No human casualties were reported, as this was an unpopulated spot in Siberia.

Sharon Begley of *Newsweek* once quoted John Pike of the Federation of American Scientists as saying that a Tunguska-sized rock from outer space could kill 70,000 people if it hit in rural American and 300,000 if it struck an urban area.[31] Maybe. Although it helps to remember that a Tunguska-sized rock did hit the Earth a century ago, and its human death toll was a nice round number: zero.

Does Tunguska have antecedents? As Gregg Easterbrook elucidated in the *Atlantic Monthly*, geophysicist Dallas Abbott of Columbia University has argued that space rocks of, respectively, 3–5 kilometers and 300 meters struck the

Indian Ocean around 2800 B.C. and the Gulf of Carpentaria in 536 A.D.[32] The latter led to poor harvests and cold summers for two years, while the former may have unleashed a planetary flood. Abbott's evidence is a crater 18 miles in diameter at the bottom of the Indian Ocean, the impact from which she believes a 600-foot-high tsunami wracked incredible devastation. It should be noted, as the *New York Times* did, that "Most astronomers doubt that any large comets or asteroids have crashed into the Earth in the last 10,000 years." Abbott and what she calls her "band of misfits" in the Holocene Impact Working Group take a decidedly minority view of the matter, and while that does not mean that they are wrong, it does mean that their alternative estimation of the frequency of 10-Megaton-size impacts — once every 1,000 or so years as opposed to the more generally accepted once every million years — should be viewed with great skepticism.[33] (Easterbrook, ignoring the majority of scientists who dispute Abbott's contentions, concludes that "Our solar system appears to be a far more dangerous place than was previously believed.")

Easterbrook is a fine science writer but his piece contains certain telltale phrases (100-kilometers asteroids are "planet killers" and NASA's asteroid and comet-hunting efforts are "underfunded") that point to an expensive conclusion. He takes up the cause of Dallas Abbott, who complains that "The NASA people don't want to believe me. They won't even listen."

Consider this quote: After noting that scientists estimate that a "dangerous" object strikes the Earth every 300,000 to one million years, Easterbrook asks William Ailor of The Aerospace Corporation, "a think tank for the Air Force," what his assessment of the risk is. Ailor's answer: "a one-in-10 chance per century."[34]

If true, humankind had better get on the case, as the survival of the species is in imminent peril. It just so happens that the Air Force is willing to take your money.

You have to hand it to the asteroid watchers: they coined a great name. Potentially Hazardous Objects (PHOs) are those with an orbit that carry them within five million miles of earth's orbit and that shine with an absolute magnitude (brightness) of at least 22, which would indicate a diameter of approximately 50 meters. There are, at present, somewhat more than 1,000 potentially hazardous objects — the typing of the name brings a shiver — according to the International Astronomical Union, which is charged with listing them. NASA estimates that about one in five NEOs — near-earth objects — is a PHO. (Acronyms are all the rage in this community.)

1997 XF11, the asteroid discussed at the beginning of this chapter, was not the only potentially hazardous object to grab the headlines. On March

4, 2004, a 20-foot-wide asteroid with the name-in-questionable-taste of FU162 floated just 4,000 miles from earth. This was the closest asteroid flyby yet recorded.

When on March 2, 2009, the asteroid 2009 DD45, approximately 100 feet across, came within 44,750 miles of earth, the media reported it this way: "An asteroid of a similar size to a rock that exploded above Siberia in 1908 with the force of a thousand atomic bombs whizzed close past Earth on Monday, astronomers said on Tuesday."[35]

Whew — *that* was a close call!

Or was it?

"Near passes" by asteroids — say, coming as close to the Earth as the moon is, which is about 238,000 miles, or 384,000 kilometers — seem to occur every few years, and are no more to be feared than is the moon. Nevertheless, the misleading phrase "near pass" gives them an aura of danger — which is the point.

A 1,000-foot wide asteroid with the insect-sounding name of Apophis is near the top of the current Hit Parade of potential Earth-killers.

Discovered in 2004, Apophis (or NEO99942), as the McClatchy Newspapers chain reported in late 2008, "has a one in 44,000 probability of slamming into our planet on Easter Sunday, April 13, 2036" — a date so full of meaning for Christians and triskaidekaphobics that it seems to have been drawn up in coincidence heaven. It has even inspired a website: www.doom2036.com, which reports on the approach of this "Continent Killer."

"We have not eliminated the threat in 2036," Lindley Johnson, head of NASA's near-earth object program, told a committee of the National Academy of Sciences. If reckoning odds at 44,000–1 isn't eliminating a threat it is a good approximation thereof, but the asteroid-hunters are determined to squeeze Apophis for every dollar. The asteroid will first pass harmlessly near Earth in 2029, an event that "could be crucial" to the planet's future — or at least crucial to the near-earth object budget.[36] In response, plans are being circulated to step up funding related to asteroid detection and defense. The B612 Foundation urged NASA to consider sending a spacecraft to the asteroid to emplace a radio transponder from which we might better determine its orbit. NASA's polite response was, in effect, that we have plenty of time, but the lesson B612 drew from the exchange and looming shadow of Doomsday 2036 was that "the responsibility for protection of the Earth from future NEO impacts should be assigned to a capable US Government agency. Such responsibility should include, inter alia, early warning capability, deflection capability and related policy development authority."[37]

That sound you hear is the Air Force, still sore over losing the space program to NASA so many decades ago, chomping at the bit, eager to take on the defense of the planet for the next 50 million years.

The deadliness of an impact depends upon several variables, among them the size, speed, makeup of the impacting body, and ground zero.

The Torino scale, its name echoing the Richter scale, was devised by Richard Binzel of MIT as a way of assessing the danger of an asteroid or comet impact. An object's Torino number is determined by its size and the probability of its collision with Earth. A zero denotes an event of "no consequence" and a 10 is a certain collision with global consequences. The devil is in the details, or in those numbers one and above. Apophis — Doomsday 2036 — is a big fat zero. The ballyhooed XF11, you will recall, was a 3.5 on the Torino Scale, or a full 3.5 units above what it ought to have been rated, so the scale still needs some fine-tuning.

In an unfortunate echo of the Department of Homeland Security's color-coded advisory system, the Torino Scale also assigns colors to the various numerical ratings. In ascending order of hazard, they are white, green, yellow, orange, and red. At present, only one known asteroid scores above zero or outside white on the Torino Scale: 2007 VK184, a 130-meter rock that is due in our celestial neighborhood between 2048 and 2057. It is rated a 1. A 1 means, according to NASA, "A routine discovery in which a pass near the Earth is predicted that poses no unusual level of danger. Current calculations show the chance of collision is extremely unlikely with no cause for public attention or public concern. New telescopic observations very likely will lead to re-assignment to Level 0."[38]

Not exactly the silver-screen-worthy stuff over which tubs of buttered popcorn are consumed. As it is, the chances of a VK184 impact are now set at one in 2,940.

Apollo astronaut Rusty Schweickart, a founder and director of the B612 Foundation, whose letterhead proclaims a desire to "Significantly alter the orbit of an asteroid in a controlled manner by 2015," puts the question in its starkest form: "Are we going to let a space strike kill millions of people before we get serious about this?"[39]

Let's hope not, but before we declare a full-out red-alert phasers-on-kill war against the Armageddon asteroid, perhaps we should ask if those who assert a far-higher-than-generally-accepted chance of a collision have a stake in the debate. The aforementioned Mr. Ailor's Air Force think tank, for instance, "seems eager to fill" the place that NASA might otherwise occupy in defending the Earth against space rocks, writes Easterbrook. "If the Air

Force won funding to build high-tech devices to fire at asteroids, this would be a major milestone in its goal of an expanded space presence."[40]

The B612 Foundation that is so eager to sic the feds on an asteroid hunt is named after the asteroid on which Antoine de Saint-Exupery's beloved fictional Little Prince lives. At least the name is warm and cuddly and not menacing in the killer asteroid vein. Founded in 2002, the group professes concern "about the current lack of action to protect the Earth from the impact of near Earth asteroids."[41] What is praiseworthy about B612 is that instead of just whining about the lack of government funding for its favored project, it seeks private funds to demonstrate the feasibility of asteroid deflection.

Founder Rusty Schweikart's preferred method is to nudge the asteroid by a "gravitational-tractor" method — in Easterbrook's formulation, landing "a spacecraft weighing only a few tons," whose presence on the asteroid is enough to push the body along a slightly different course.[42] Schweikart and B612 propose to rendezvous and dock with a near Earth asteroid and then "push it (gently and for a long time)" into a different orbit. The time-frame: by the year 2015. This is an arbitrary date, but the B612 activists understand that unless you attach a deadline to a project, it's not going to get done.

The B612 Foundation, not surprisingly, accepts the most "optimistic" (if that word is appropriate) estimates of the impact community, asserting that there is a 2 percent chance of an "unacceptable collision" in the next century. Impact expert Clark Chapman, it should be noted, is on the board of directors, and Chapman is not exactly a killer-asteroid skeptic. And the foundation is not exempt from the scaremongering rhetoric of the field. "[W]e know," its website declares, "that it is only a matter of time before we detect a NEA headed for the Earth, and yet we have done nothing to prepare for it." Well, true, in a way, if you discount the by now numerous workshops and conferences on NEAs, the first of which was authorized by Congress way back in 1990, before *Deep Impact*, in those days when Bruce Willis had hair and Demi Moore. But more to the point: the "matter of time" doesn't mean it will happen at some point in the next few months, or years, or decades, or centuries — it could well be millennia before a NEA of any significant size, even Tunguska-like, appears. We might be forgiven for not exhibiting a state of urgency.

"We understand that sooner or later our survival, not just as individuals, but also as a species, will depend on being able to dissuade one of these cosmic neighbors from dropping in uninvited," states the B612 website. Again, the devil is in the details, or more specifically, in the quite possibly millions of years that are contained in the vernacular phrase "sooner or later." Sooner or later the Sun will burn out, too — but must we mobilize against that remote possibility at this time?

"The public policy reality," states the B612 Foundation, "is that when the scientific community makes an announcement that a NEA is headed toward an impact with Earth, the public will expect that something has already been done to prepare for this eventuality. The horror is that, at this time, nothing of this kind is being done!"[43]

The horror? A slight overstatement, it seems.

Detect. Deflect. Or intercept. These are the stages of Earth defense that have been the subject of NASA workshops on "the detection and interception of NEOs," one of which (the NASA Near-Earth Object Detection Workshop) recommended a "Spaceguard Survey," whose network of six 2.5 meter telescopes would scan the skies for NEOs and keep gainfully employed a segment of the astronomy community.[44]

Writing in *Nature*, Thomas J. Ahrens of the Seismological Laboratory at the California Institute of Technology and Alan W. Harris of the Jet Propulsion Laboratory examined ways to divert, deflect, or demolish Earth-bound asteroids and comets.

A small asteroid could be deflected, they say, by the "direct impact of a spacecraft-borne mass." The force of the impact would create ejecta, which would perturb the body's orbit. For larger bodies, the current strategy of choice would be to explode a nuclear device above the surface of the body; the "subsequent blow-off of surface material" would perturb the orbit and deflect the potential killer from Earth. Of course, the defenders of Earth could explode the nuclear device on the asteroid's surface, which "effectively induces a velocity change" and saves Earth, too.[45] Unfortunately, there is no obvious way to keep fragments of the exploded asteroid from embarking on a course whose terminus might include the Earth. Countless smaller chunks would jet out from the destroyed body, and where they would go is anyone's guess. The "practical problems of dealing with the numerous, randomly deployed fragments of a disrupted body would be enormous," write another team of researchers without understatement.[46] Thus "asteroid deflection, rather than destruction by fragmentation, appears to be the most efficient strategy."[47] (The exception would be for very small bodies, which could be safely pulverized.)

This is a tentative conclusion, with an emphasis on *tentative*: the problem it seeks to solve may not present itself for another 50 million years. But for now, as a team of researchers from the Russian Academy of Sciences determined, destroying a killer asteroid with nuclear charges "presents an almost certainly insuperable technological problem."[48]

Because the "technologies that might be employed to divert asteroids can be expected to change so rapidly in the coming decades," write Ahrens and Harris, "it would be premature to conduct detailed engineering studies or to build prototypes at this stage." The aerospace–industrial complex might demur, but "the low probability of impact of hazardous asteroids, the high cost in the face of a low risk factor, and the rapid changes that are to be expected in defense systems technology" combine to make any significant expenditure on asteroid deflection programs at best unwise, at worst a colossal waste of money.[49] The current technologies will be overtaken by new ones so quickly as to make any defense system almost instantly obsolete.

Moreover, many astronomers warn against letting fear of an exceedingly remote possibility stampede us into amassing an arsenal to fight the nonexistent threat. Says Jay Melosh, professor of planetary science at the University of Arizona, "Maintenance of a large fleet of gigaton-class nuclear weapons is probably more dangerous to humanity than the threat from which they are supposed to defend us."[50]

It should be noted that not all scientists dismiss nuclear detonations as wholly inappropriate for asteroid disposal. Four Russian scientists from the Institute for Technical Physics in Chelyabinsk and Johndale C. Solem of the Los Alamos National Laboratory believe that "astral assailants"[51] — a good alliterative phrase — could be demolished by nuclear explosive devices with a greater than 1 Mt yield. Detonated just above the surface of the AA — astral assailant, that is — these Earth-saving nukes would need to be powerful enough — with yields up to 100 Megatons — to "shatter the object into small fragments that will be unable to penetrate the Earth's atmosphere."[52] But this is a minority view.

"Trajectory modification" can be achieved by applying energy to the asteroid, whether by explosions or the landing of a foreign object, which is presumably delivering a payload. The problem with the latter, as Joseph G. Gurley, William J. Dixon, and Hans F. Meissenger write in "Vehicle Systems for Missions to Protect the Earth Against NEO Impacts," is that, once again, technological change is so rapid that developing an NEO-landing craft would be ridiculously premature. Rather, "major development costs" should be "delayed until an actual threat is detected."[53] This is not exactly music to the ears of the NEO crowd.

Even deflection of a NEO entangles the parties involved in what Alan W. Harris and Steven J. Ostro of Jet Propulsion Laboratory, Gregory H. Canavan of the Los Alamos National Laboratory, and Carl Sagan term "the deflection dilemma." For the technology could also be used by malicious rulers or terrorists to "deflect a nonmenacing NEO so it impacts the Earth."[54] The prospect cries out for a James Bond film.

Given that there "is no known incident of a major crater-forming impact in recorded human history," argues P.R. Weissman of the Jet Propulsion Laboratory, and since "the credibility of the impact hazard" is justifiably low with the public and governmental decision-makers, we ought to defer the development of a defensive system until such time as technological advances permit us to do so at a reasonable cost.[55]

There is also, he points out — at the risk of being called chauvinist, no doubt, by the more feverish Earth-savers — the "pragmatic and/or parochial" fact that the United States accounts for 6.4 percent of the total land mass of the Earth, and only 1.9 percent of the total area, including water.[56] Thus anything short of a civilization-ending asteroid would be exceedingly unlikely to hit the U.S. By contrast, such threats as infectious diseases and nuclear war present a more real and immediate danger to Americans, and to earthlings in general. Perhaps money would be better spent addressing those matters?

Absent a panic, the lack of any popular support for "Earth defense" vexes the Earth-savers. One lesson of human history, say Robert L. Park of the American Physical Society, Lori B. Garver of the National Space Society, and Terry Dawson of the U.S. House of Representatives, is that "societies will not sustain indefinitely a defense against an infrequent and unpredictable threat." There is almost no popular constituency for asteroid defense, and it is sheer hubris to believe that any defense we are capable of designing today will be of anything more than "historical interest" to our descendants a century or a millennium hence.[57] We are not the alpha and the omega. Our most sophisticated weapons will be to our distant descendants as spears are to us.

As for defense against comets, the lesser-discussed threat from the skies, they are "astonishingly intractable," write Clark R. Chapman, Daniel D. Durda, and Robert E. Gold.[58] Hard to find, their motions difficult to predict, their structure and consistency questionable, comets are the wild card in the deck of Earth-killers, though since one hasn't hit us in at least many tens of millions of years, only the most hopeless superlunary hypochondriac is going to lose much sleep over them.

For a near-impossible scenario, an awful lot of laser ink has gone into studies of the consequences of an impact. Let's face it: The topic is sexy.

The effects of an Earth-space rock collision with energies below 10 Megatons would be "negligible," write Owen B. Toon, Kevin Zahnle, and David

Morrison of the NASA Ames Research Center, Richard P. Turco of UCLA, and Curt Covey of the Lawrence Livermore National Laboratory, in *Reviews of Geophysics*. Impacts measuring between 10 Megatons and 10 to the 4th power Megatons — say, comets and asteroids with diameters of less than 400 meters and 650 meters, respectively — would be equivalent "to many natural disasters of recent history." In other words, death-dealing but manageable in a global sense. Those with an energy range in the 10 to the 5th–6th power Megatons are "transitional" — the fires, earthquakes, and tsunamis would unleash devastation, though the authors do not believe a "global catastrophe" would occur at less than an energy level of 10 to the 6th power Megatons. They do admit to "considerable uncertainty," noting that previous estimates may have overstated the damage at certain levels of impact, though they say, with great wisdom, that "it is to be hoped that no large-scale terrestrial experiments occur to shed light on our theoretical oversights."[59]

They can say that again.

The impact upon the Earth of an object of more than 400 meters in diameter crashing into an ocean would be a tsunami, an enormous wave created by the impact of the asteroid or comet upon the ocean floor, which could cause massive numbers of deaths due to drowning, though it would be highly unlikely to cause extinction of the human species. A wall of water — a wave over 60 meters high — would sweep over the impacted ocean's coasts. The huge and widespread fires would claim uncounted lives, too, and the "opacity of the smoke generated by the fires" would contribute to the sharply reduced level of sunlight upon the Earth.

The consequences of an impact with an energy of 10 to the 7th power Megatons could be K–T like, as 100-meters-high tsunamis swamp coastal zones, fires rage around the world, and "Light levels may drop so low from the smoke, dust, and sulfate as to make vision impossible."[60] Photosynthesis, too, becomes impossible, and food supplies disappear. Dwellers in sea and on land perish of fire, starvation, or flood. In the aftermath, survivors would compete with rodents for the available food. (As paleontologists Peter M. Sheehan and Dale A. Russell note, "In the short term domestic cats might play a useful role in protecting food supplies."[61] Humans, they believe, would survive such a catastrophe, though in greatly reduced numbers and for millennia they would be vegetarians practicing subsistence agriculture. No doubt, that sounds appealing to some of the greener readers.)

If an impact with a smaller body is sometimes compared to the aftermath of a nuclear war, the fact that in a war the civilian infrastructure is specifically targeted means that it is "much more likely that society could cope with the problems following a small impact better than it could adjust to the problems following a nuclear war," according to Toon, Zahnle, et al.[62]

Interestingly, the authors say that acid rain — very much a fashionable environmental cause in the 1980s, though it has since receded before global warming — would not be a widespread problem, although the rain may well be acidified due to the nitric oxide resulting from impact-induced shock waves.

We should here acknowledge, without necessarily casting aspersions on any of the papers discussed in this chapter, the tendency of scientific journals to publish sexy articles. (Sexy, at least, by the decidedly unsexy standards of scientific journals.)

Writing in the *Public Library of Science*, Neal S. Young of the National Institutes of Health, John P.A. Ioannidis of the Biomedical Research Institute in Greece, and Omar Al-Ubaydli of George Mason University applied what economists call the "winner's curse" of auction theory to scientific publishing. Just as the winner in, say, an auction of oil drilling rights is the firm that has made the highest estimation — often overestimation — of a reserve's size and capacity, so those papers that are selected for publication in the elite journals of science are often those with the most "extreme, spectacular results."[63] These papers may make headlines in the mainstream press, which leads to greater political pressure to fund projects and programs congruent with these extreme findings. As *The Economist* put it in an article presenting the argument of Young, Ioannidis, and Al-Ubaydli, "Hundreds of thousands of scientific researchers are hired, promoted and funded according not only to how much work they produce, but also where it gets published." Column inches in journals such as *Nature* and *Science* are coveted; authors understand full well that studies with spectacular results are more likely to be published than are those that will not lead to a wire story.

The problem, though, is that these flashy papers with dramatic results often "turn out to be false."[64] In a 2005 paper in the *Journal of the American Medical Association*, Dr. Ioannidis found that "of the 49 most-cited papers on the effectiveness of medical interventions, published in highly visible journals in 1990–2004... a quarter of the randomised trials and five of six non-randomised studies had already been contradicted or found to have been exaggerated by 2005."

Thus, those who pay the price of the winner's curse in scientific research are those, whether sick patients or beggared taxpayers, who are forced to either submit to or fund specious science, medical or otherwise. The trio of authors call the implications of this finding "dire," pointing to a 2008 paper in the *New England Journal of Medicine* showing that "almost all trials" of anti-depressant medicines that had had positive results had been published, while almost all trials of anti-depressants that had come up with negative results "remained either unpublished or were published with the results presented so that they would appear 'positive.'"

Young, Ioannidis, and Al-Ubaydli conclude that "science is hard work with limited rewards and only occasional successes. Its interest and importance should speak for themselves, without hyperbole." Elite journals, conscious of the need to attract attention and stay relevant, cutting edge, and avoid the curse of stodginess, are prone to publish gross exaggeration and findings of dubious merit. When lawmakers and grant-givers take their cues from these journals, as they do, those tax dollars ostensibly devoted to the pursuit of pure science and the application of scientific research are diverted down unprofitable, even impossible channels. The charlatans make names for themselves, projects of questionable merit grow fat on the public purse, and the disconnect between what is real and what subsidy-seekers tell us is real gets ever wider.[65]

The matter, or manipulation, of odds in regards to a collision between a space rock and Earth would do Jimmy the Greek proud. As Michael B. Gerrard writes in *Risk Analysis* in an article assessing the relative allocation of public funds to hazardous waste site cleanup and protection against killer comets and asteroids, "Asteroids and comets are... the ultimate example of a low-probability/high-consequence event: no one in recorded human history is confirmed to have ever died from one." Gerrard writes that "several billion people" will die as the result of an impact "at some time in the coming half million years," although that half-million year time-frame is considerably shorter than the generally accepted extinction-event period.[66]

The expected deaths from a collision with an asteroid of, say, one kilometer or more in diameter are so huge that by jacking up the tiny possibility of such an event even a little bit the annual death rate of this never-before-experienced disaster exceeds deaths in plane crashes, earthquakes, and other actual real live dangers.

Death rates from outlandish or unusual causes are fairly steady across the years. About 120 Americans die in airplane crashes annually, and about 90 more die of lightning strikes. Perhaps five might die in garage-door opener accidents. The total number of deaths in any given year by asteroid or meteor impact is zero — holding constant since the dawn of recorded time.

The numbers from the first three causes fluctuate somewhat from year to year; again, the last number has been zero since the beginning of recorded history. The chances that in a given year an American will die as a result of being struck by lightning are about 100 in 300 million, or 1 in 3 million. Extraordinarily remote, though it is still a good idea to seek shelter in a storm. And what of the chances that an American will die in an asteroid or comet crash? This is far harder to calculate. Do such impacts happen every 65 million years? If so, the chances are so minuscule as to be off the charts. Or are we overdue? Does a Tunguska-type impact happen every century? (Other than at Tunguska, it hasn't in recorded history, but again, maybe we're overdue.)

Michael B. Gerrard, in *Risk Analysis*, uses the highly questionable assumption that four billion people die due to such an impact every one million years to come up with an annual asteroid/comet death rate of 4,000 persons per year. (Remember, the actual number in all of human history is zero.) Asserting that conventional financial analysis places the value of a human life at $4–8 billion, Gerrard writes that at "that rate, comet and asteroid detection would warrant spending $16–32 billion per year," or far more than the measly millions the governments of the world devote to that purpose today.[67]

But there is hope! Gerrard writes that elements of the space and defense industries "see NEO defense as a post-Soviet substitute for President Reagan's Strategic Defense Initiative ('Star Wars')" and that these "industries have begun to exert their influence to increase federal expenditures for NEO related projects."[68] Armed with optimistic (or, from the point of view of humankind, extremely pessimistic) estimates of the frequency of asteroid/comet impacts, they will prove a formidable lobby, made all the more tenacious by the open-ended — in fact, infinite — horizon for the project of saving the Earth.

In the Tom Gehrels-edited volume *Hazards Due to Comets and Asteroids*, a foursome of researchers affiliated with, among institutions, the U.S. Air Force Phillips Laboratory and the Ballistic Missile Defense Organization (formerly the Strategic Defense Initiative Organization), explore how SDI technology might be adapted to the needs of the "civilian space community."[69] Note that "civilian" is not a synonym for "private" but rather non-military governmental. If, indeed, NASA can be considered non-military. The feeding frenzy is on.

We pause here for one of the great lines in the history of Fund Me or the World Will End-ism: "The dinosaurs became extinct because they didn't have a space program. And if we become extinct because we don't have a space program, it'll serve us right!"[70] The science-fiction author Larry Niven coined that one, and it serves as an exemplary example of the genre.

The number of asteroid-detection projects expands with the hype. Tom Gehrels, an astronomer at the University of Arizona, founded Spacewatch at the Kitt Peak Observatory in Tucson. In 1989, Spacewatch turned one of the observatory's venerable old telescopes to the search for near-earth asteroids. It continues the watch about twenty nights per month and has augmented it with a larger scope as well. Spacewatch accepted private donations but received federal funding from NASA and the U.S. Air Force Office of Scientific Research. Gehrels has a knack for the doomsday quote. For instance: "There are some 1,700 1-kilometer or larger asteroids out there, and one or two or so might have our name on them. That is an intolerable uncertainty that we, as scientists, are responsible for and have to take care of. The probabilities [of an impact] are low... but the consequences are horrendous."[71]

True, they are horrendous, though "low" may be understating the infrequency of an event that may happen once every 50 million years or so.

Other such programs include NEAT (Near-Earth Asteroid Tracking system), a NASA and U.S. Air Force-subsidized project whose central instrument is the 39-inch telescope atop Mt. Haleakala on Maui and whose investigating team is based at the Jet Propulsion Laboratory; and LINEAR (Lincoln Near-Earth Asteroid Research project), which was a kind of spin-off from Reagan's Star Wars missile defense research. The Air Force, it seems, is winning easily whatever intraservice rivalry may have been spurred by the doomsday crowd.

Project Spaceguard, its title taken from an Arthur C. Clarke novel, grew out of the cooperation of NASA and international agencies interested in the potential problem of NEOs. In fact, Clarke, before he died in 2008, took what 999 of 1,000 people might consider *too* long a view in deciding what to fund. He said, "We need to look to survival strategies for 3001, not just 2001."[72] In this view, the funding possibilities become endless: a study of traffic patterns for human-jetpack highways; or perhaps the composition of a Bill of Rights for robots; maybe a guide to public restroom etiquette for transhumans.

Duncan Steel was one of 24 members of the Spaceguard Committee, a NEO project which "comprised a global network of at least six dedicated search telescopes, three each in the northern and southern hemispheres." The proposed quarry? Asteroids of at least one kilometer. Though its cost was estimated at a relatively modest $300 million over 20 years — somehow one thinks that it would have gone higher — the U.S. was the only nation that signed on. Spaceguard was stalled. So in 1996 astronomers from several nations formed the Rome-based Spaceguard Foundation, which lobbies for NEO programs. Steel directed Australia's NEO detection and tracking program from 1990 to 1996, he tells us, before it was cancelled "despite fervent lobbying by virtually the entire world community of scientists involved in

Spaceguard." The idiots who cancelled Steel's mission didn't understand that "we are trying to safeguard the future of the human race."[73] And you wanna argue dollars and cents?

The evolutionary development of mathematical ability, says Duncan Steel, equips us to avoid the fate of the dinosaurs. For the Earth story has a happy ending. "Eventually a species arose with the capability to take its destiny into its own hands, and ensure that its days are not brought to a premature end by some rogue asteroid or doomsday comet. We are that species."[74] Or at least we are if we fund Spaceguard.

In their foreword to Duncan Steel's *Target Earth*, Brian Marsden, director of the Minor Planet Center at the Harvard-Smithsonian Center for Astrophysics and vice president of the Spaceguard Foundation, and Dr. Andrea Carusi, president of the Rome-based Spaceguard Foundation, state that "there is a greater chance of an individual (you, reading this book) dying through an asteroid striking our planet than in a jetliner crash." Doubtful, but go on. Spaceguard, therefore, is simply a "cosmic insurance policy," and if its enrichment happens to redound to the benefit of the authors, well, so much the better. After all, "our planet is indeed a target in a celestial shooting gallery" — albeit a gallery in infinite space in which the shooters are blind.

They conclude this threatening foreword, "If humans are to continue to thrive on Earth, and perhaps eventually expand into space, we must avoid being hit by some piece of cosmic detritus that could put an end to civilization."[75] Well, yes.

Duncan Steel makes Marsden and Carusi look positively complacent. His world is under potential assault by "lumps of rock, metal, and ice" which "pose a substantial and surprising hazard to the well being of each and every one of us." Our condition is "frightening," it is "scary," and you had best be on your toes. "We live not on Planet Earth," he informs the reader, "but on Target Earth."[76]

Does this sound alarmist? Good! "Perhaps there is need for a little cosmophobia — a realization that all is not benign out there in space." They're out there — waiting, watching, plotting extinction in the deepest recesses of the Oort Cloud.

(With chapter headings such as "Mars Attacked" and "Target Earth," Steel's book has a fun kind of B-movie '50s science-fiction feel, though when considered as a piece of special-pleading for governments funds, the fun wears off.)

A NEO program is "a bit like an insurance policy," says Steel, reassuringly. "You don't insure your car and then at the end of the year bemoan the fact that you've had no accidents." Only a fool carries no car insurance. After all, once found, asteroids can be tracked in their orbits and monitored in the way that police might monitor potentially troublesome characters — if those characters' movements followed a predictable arc, that is.

"If we are to have a long-term future on planet Earth," declares Steel, "then our increasing knowledge of the projectiles that cross our Sun-centered orbit has made one thing clear: we must learn to dominate the solar system."[77] This sounds rather martial, but he's only getting warmed up. His funding source dried up, Steel uses every dire image he can find to make his point. In discussing Hermes, an asteroid of about half a mile in diameter, which came within twice the distance of the moon to the earth in 1937 and then drifted off into supernal oblivion, Steel writes that "it could hit the Earth at virtually any time, without warning." Yes, and men might sprout wings tomorrow, too, but we needn't lose any sleep over the possibility.

"When you are staring down a loaded gun barrel, there's nothing you won't do to save your life," he writes of averting a collision. True. "Similarly, no expense is too great to avoid asteroidal Armageddon." So shouldn't those searching out Armageddon-bringing asteroids in the sky be good candidates for public subvention? He ends with a threat in the form of a question: "I'm sure of one thing. To have a fighting chance of avoiding an impact catastrophe set up by the clockwork of the heavens, we have to find our enemy quickly. [Notice the language of war. He has figured out that military and defense gets funded lavishly. The National Education Association and the National Science Foundation have nothing on him.] There is no excuse for not carrying out the necessary space surveillance program. We must have an answer to that vital question: Is there a big one due to hit us soon?"[78]

For most of the 1990s, NASA devoted about $1.5 million per annum to NEO detection. That figure doubled in 1998, the year NASA undertook Spaceguard to track near Earth objects of at least 1 kilometer in diameter. It was budgeted at $4.1 million annually from FY 2006 to FY 2012. As this book went to press, over 700, or about 70 percent of the estimated total of perhaps 1,000–1,200 1-km NEOs, have been located. None is on a collision course with Earth. Going down in size, there may be as many as 100,000 NEOs of more than 100 meters but less than 1 kilometer. Scientists have discovered perhaps 10 percent of them, so their detection and tracking would provide a nice living for a whole academy's worth of astronomers. NASA's Donald Yeomans expresses the hope that the next generation of researchers will discover 90 percent of the potentially hazardous objects of at least 140 meters in diameter.[79]

Not coincidentally, 1998, the year that Spaceguard's funding spiked upward, was the year in which both *Deep Impact* and *Armageddon* were released. Who says movies don't drive public policy?

"[P]ro-active steps" to combat this menace are the recommendation of Clark R. Chapman and Daniel D. Durda of the Southwest Research Institute and Robert E. Gold of Johns Hopkins University. In *The Comet/Asteroid Impact Hazard: A Systems Approach* (2001), Chapman and his coauthors used what has become the reigning cliché in the field: "The dinosaurs," they said, "could not evaluate and mitigate the natural forces that exterminated them, but human beings have the intelligence to do so." To do nothing — to be inactive rather than proactive — "could conceivably doom everyone we know and everything we care about."[80] You don't want *that* on your conscience, do you?

The authors ask whether or not the "NEO hazard" has been "overblown" by Hollywood hype. They see, in 2001, a political climate of reduced government spending (in fact it zoomed under President Bush) and "massive tax cuts" (when? where? for whom?). They concede that many sensible scientists argue that technological advances are so rapid and unpredictable that it is foolish to design a NEO-deflector today when it may not be needed until some future date when our contemporary science would seem like so many primitive sticks and stones.[81] Nevertheless, Chapman and his coauthors regard the expenditure of government monies on comet/asteroid defense as wise and necessary.

Chapman has been perhaps the most prolific author of the subject. In his paper "How a Near-Earth Object Might Affect Society," presented to a 2003 conference on NEOs in Frascati, Italy, he lays out six impact scenarios, which he prefaces with the prediction that "the increasing rate of discoveries of Near Earth Asteroids combined with media sensationalism" will put NEOs into the Action basket on the desks of "responsible emergency management officials." No doubt this is true, especially as the media report every distant passage of an asteroid as a "near miss," fanning the flames of the doomsday industry.

Chapman de-emphasizes the costs of impact preparation, arguing that since the physical manifestations of a collision — tsunami, earthquakes, fires — are already part of existing disaster-mitigation efforts, what would be needed are mostly "relatively low-cost, marginal add-ons" to civil defense programs.

The smallest falling bodies, those with diameters under a few meters, are of "no practical concern," says Chapman, and in fact they are to be desired, at least by those who keep their eyes on the skies watching for brilliant fireballs whose burning up in the atmosphere provides a show far more spectacular than the most lavish Fourth of July fireworks.

Even bodies with diameters of 10–30 meters, of which Chapman estimates six may fall to earth in a century, cause little more than broken windows. They explode too high in the atmosphere to cause serious harm. The next largest potential strikers of Earth are those in the Tunguska range of 30 meters–100 meters. The shock waves from the atmospheric explosion would "topple trees, wooden structures and ignit[e] fires within 10 kilometers," writes Chapman. Human deaths could result if the explosion took place over a populated area. Though Chapman estimates the likelihood of a Tunguska occurring in any given century at four in ten, it is worth noting that there is no evidence that such an explosion has killed a single human being in all of recorded history. Either we're overdue or that 40 percent is high. Moreover, given that the location of such an explosion is utterly unpredictable, it would be far more likely to happen over an ocean or a desert than over, say, Tokyo or Manhattan. The after effects would be minimal, and Chapman says that "nothing practical can be done about this modest hazard other than to clean up after the event." In fact, "It makes no sense to plan ahead for such a modest disaster... other than educating the public about the possibility." The cost of a telescopic survey capable of picking up bodies of such diminutive size would be prohibitive. It would be the ultimate Astronomers Full Employment Act.

A body of 100 meters–300 meters in diameter would either explode at low altitude or upon impact with the ground; it would be "regionally devastating," but Chapman pegs the chances of such a catastrophe at 1 percent per century. A small nation could be destroyed by the impact of a body of 300 meters—1 km in diameter, or a "flying mountain" of sorts, which would explode with energy yield ten times more than "the largest thermonuclear bomb ever tested." If striking land, it would carve out a crater deeper than the Grand Canyon. If it hit a populated area, the death toll could be in the hundreds of thousands. The likelihood of such a collision Chapman estimates at 0.2 percent per century. An asteroid or comet of 1–3 kilometers in diameter would cause "major regional destruction," possibly verging on "civilization-destruction level." Chapman puts the chances of this at 0.02 percent per century. The impact of a body more than 3 kilometers in diameter might plunge the Earth into a new Dark Age, killing most of its inhabitants, though the chances of this are "extremely remote" — less than one in 50,000 per century. Finally, mass extinction would likely occur should a body greater than 10 kilometers pay us a visit, though the chances of this are less than one in a million every century, or so infinitesimal that even the most worry-wracked hypochondriac will not lose sleep over the possibility.

In fact, for any impact with a Chapman-calculated likelihood of less than one in a thousand per century, he concedes that there is "little justification for mounting asteroid-specific mitigation measures." The chance of a civilization-ender is so remote that he counsels no "advance preparations" — or almost none. For Chapman recommends further study of NEOs, as well as investigation into methods of their diversion.[82] This is exactly what the NEO lobby wants.

Perception is critical to funding, as advocates of NEO programs as well as more fanciful Earth protection schemes understand. Writing in the massive compendium *Hazards Due to Comets and Asteroids*, Chapman, David Morrison, and Paul Slovic asked whether, in the absence of specific and concrete predictions of a collision, the public concern would "be great enough to induce them to support expenditures of public funds to detect threatening asteroids or comets?"[83]

Evidence piled up on both sides of the question. For the negative, they cited studies showing that people fear technological hazards more than natural disasters, especially natural disasters that have an extremely remote chance of occurring. On the other hand, the asteroid risk is "demonstrable" (if the Alvarez theory is to be credited) and while the odds are astronomical, the consequences are eschatological.

Coauthor Slovic supervised a survey of attitudes toward the impact hazard among students at the University of Oregon. This was conducted shortly after a 1992 *Newsweek* cover story on killers from outer space and it found that students ranked "asteroid hitting the Earth" as the 14th most serious hazard out of a list of 24. It ranked behind hurricanes and earthquakes but above X rays, floods, prescription drugs, bacteria in food, and, surprisingly, air travel. When told that "Scientists say that a civilization-threatening asteroid impact can be expected every 300,000–1,000,000 years" — an optimistic/pessimistic (it depends on whether you are putting money in or want to take it out of the treasury) overestimate — 56 percent then agreed that "we don't really have to worry about this threat in our own lifetimes." One wonders if the other 44 percent have discovered an anti-aging pill.

The students chose prudence over panic. They expressed "modest support" for NEO detection but "considerable opposition" to putting weapons into space to meet this phantom threat. "Support for asteroid tracking and defense systems was greatest among those who tended to trust both the scientific community and the government" — the credulous, who are always the first to support any government program putatively designed to address a "crisis."[84]

The authors conclude that "increasing public awareness" may be the key to "demands... for action to deal with the impact hazard." And so for the fifteen years since the Gehrels volume was published, those who seek public funds to deal with this rather hazy "hazard" have done their best to raise public awareness through various means.[85]

Curiously, a bracing shot of skeptical clarity appeared in the toe-breakingly-if-you-drop-it long *Hazards Due to Comets & Asteroids* on page 1203, which one seriously doubts one in a hundred readers ever make it to. P.R. Weissman of the Jet Propulsion Laboratory writes: "One problem for those advocating an impact hazard defense and/or detection system is that their recommendations often appear to be self-serving. Astronomers who study small bodies have advocated an observing program that emphasizes searching for large (> 1 kilometer) Earth-crossing asteroids and comets.... These are, in general, the same objects that those astronomers are currently discovering with their existing search programs. Thus, their recommendations can be viewed as motivated by a desire to obtain additional funding and instrumentation for their ongoing work."[86]

What a world of wisdom and insight is contained in those sentences! Astronomers and engineers whose livelihoods depend on the perception of an impact hazard develop and publish studies concluding that there is an impact hazard. The circle goes round and round, and fills, gradually, with taxpayer money.

The lessons of Chicken Little, suggests Weissman, are one factor that has kept the funding of such programs from really taking off. Weissman urges his colleagues to "GO SLOW." Don't "attempt to divert substantial resources" to a program that, at present, is "neither necessary, nor prudent."[87]

As if to confirm Weissman, the very next paper, by William K. Hartmann of the Planetary Science Institute and Andrei Sokolov of Moscow, argues that the asteroid threat is really an "asteroid opportunity." Rather than emphasizing defense against killer asteroids, the authors say, why not spend funds "cataloging, visiting, and processing materials of NEOs"?[88] They are rich in minerals, after all — nickel, cobalt, platinum, iron. Instead of calling a NEO survey "Spaceguard," which is a defensive, even fearful name, and the product of a mistaken "mindset," write Hartmann and Sokolov, why not accentuate the positive? Asteroids should be our friends: they can be studied, mined, even made profitable. Otherwise, the authors fear, the NEO survey will look like "a boondoggle for asteroid-minded astronomers."[89]

Exactly.

That same volume contained a paper by Andrew F. Cheng and Robert W. Farquhar of the Applied Physics Laboratory, J. Veverka of Cornell University, and C. Pilcher of NASA arguing for space missions to NEOs in order to better determine their composition and structure.[90] Going them one better were a sextet of researchers from NASA, the Russian Space Agency, and other institutions who wrote of manned exploration of NEOs, which would, among other things, "strengthen the integrity of any foreseeable program of human lunar and Mars exploration."[91] The NEO scare has many spinoffs, it seems: its missions can be even shake-down cruises for that long-delayed manned mission to Mars. The challenges of dealing with microgravity and cosmic rays, designing effective life support, and ensuring communications across the void of space would be rehearsed in manned NEO missions in preparation for a trip to Mars.

Political scientist John Mueller, author of *Overblown: How Politicians and the Terrorism Industry Inflate National Security Threats, and Why We Believe Them* (2006), quotes a Turkish proverb — "If your enemy be an ant, imagine him to be an elephant" — which he describes as "spectacularly bad advice."[92] To drastically misestimate your enemy is to badly misallocate resources.

For instance, the Department of Homeland Security announces, "Today's terrorists can strike at any place, at any time, and with virtually any weapon."[93] This language, preposterously untrue — can terrorists really strike in Baton Rouge with an antimatter gun, or in Yankton, South Dakota, with a secret decoder ring? — could easily be transferred into any scare-mongering story about killer comets or rogue asteroids threatening the earth.

Carl Sagan said that the extraordinarily remote chance that an asteroid or comet might strike Earth justified — indeed, compelled — a vigorous space program. "Since, in the long run, every planetary society will be endangered by impacts from space, every surviving civilization is obliged to become spacefaring — not because of exploratory or romantic zeal, but for the most practical reason imaginable: staying alive."[94]

Now, given that a K–T like disaster is expected every 50 or 100 million years, we might be excused for asking just how imperative it is that our government lead us into the brave new world of "spacefaring." To ask this question, however, is to reveal oneself as unimaginative, dull, lacking in foresight, perhaps troglodytic, certainly far from *au courant*. This is asteroid alarmism as a trick shot, as a bulked-up NEO-detection program leads to an enhanced manned space travel program.

Just as "professors learned to hustle in the regime of largesse after Sputnik," in Walter A. McDougall's phrase, so did they prove fast on their feet in tracking down killer asteroid funds.[95] Warning, in grave sepulchral tones, about the end of the world does tend to concentrate the attention of the listener. And if the Cassandra giving the warning has a Ph.D. after her name, all the better. Surely no doctor of philosophy would exaggerate in order to have a pet project funded!

The press does its part, as it always has. Sensationalism sells, and if it isn't exactly grounded in truth, well, wink wink, everyone knows you can't always believe what you read or hear.

As usual, the middlebrow hysterics at *Newsweek*, who can spot a phony story or ridiculous trend a thousand miles away (an epidemic of crack babies! dangerous militiamen terrorizing Middle America!), took the story and ran with it. They ran with it so erratically and irresponsibly that after reading the cover story "The Science of Doom" in the November 23, 1992, issue, all one could say is, "This is even worse than *Time!*"

The author of *Newsweek*'s doomsday story, Sharon Begley, inserts into her lede the magazine's characteristic and annoying plucked-from-the-headlines phraseology, calling the demon comet a "Scud from hell," thus neatly conflating manmade and very real weapons with a fantastical threat which is so remote as to be almost unfathomable. We live, she says, in — you guessed it — a "cosmic shooting gallery." A comet six miles in length ushered the dinosaurs off the Earth, she notes, neatly segueing into a prediction by the ever-quotable Brian Marsden of the Harvard-Smithsonian Center for Astrophysics that Comet Swift-Tuttle, which by a nice coincidence also measures six miles across, has a "1-in-10,000 chance of hitting Earth on Aug. 14, 2126."

Pretty scary, huh? Also total nonsense, though the nice thing about writing for a newsweekly is that its pages are so disposable, its news so ephemeral, that you never really have to make corrections.

In fact, Comet Swift-Tuttle, which was discovered in 1862 by the American astronomers Lewis Swift and Horace Tuttle, is not going to render our descendants as dead as the dodo in 2126. Comet Swift-Tuttle made its first reappearance since the Lincoln administration 130 years later, when on November 7, 1992, it passed about 110 million miles from Earth. Brian Marsden, working from the available estimates of the comet's elongated orbit, predicted a very small chance that it could occupy the same space and time as the Earth on its next approach, 134 years hence. In Begley's words, "Ever since, the doomsday faxes and E-mail have been flying at the speed of light."

Cooler heads intervened. Donald Yeomans of the Jet Propulsion Laboratory said, "The comet will pass no closer to the Earth than 60 lunar distances [14 million miles] on August 5, 2126. There is no evidence for a threat from Swift-Tuttle in 2126 nor from any other known comet or asteroid in the next 200 years."[96] Even Brian Marsden concurred. He retracted his prediction, though he held out the possibility that in the year 3034 the comet could come within a million miles of Earth. Surveying this very false and very loud alarm, Sally Stephens, writing in the journal of the *Astronomical Society of the Pacific*, observed, "Marsden's prediction, and later retraction, of a possible collision between the Earth and the comet highlight the fact that we will most likely have century-long warnings of any potential collision, based on calculations of orbits of known and newly discovered asteroids and comets. Plenty of time to decide what to do."[97]

This last sober judgment — "Plenty of time to decide what to do" — is exactly *not* the kind of message that *Newsweek* likes to convey.

Having raised the caution flag on Swift-Tuttle despite an utter lack of evidence, Begley then asks the question that the doomsday crowd never tires of asking: "can science save the planet Earth?" Or to put it more directly, can government-funded science save the planet Earth?

The Spaceguard project by which a gaggle of telescopes would discover and monitor near-earth objects was under discussion at the time, and Begley called it a "4-cents-per-American insurance policy," leaving no doubt about which side of the issue she and any other American sentient enough to step out of the way of falling objects stood. Who could possibly begrudge Washington four cents — even penny candy costs more these days, for goodness sake! — as an "insurance policy" against the destruction of civilization, life as we know it, and our dear Mother Earth?

Begley goes on to describe the consequences of collisions with heavenly bodies up to and including "mountain size" (""would surely rank as the greatest catastrophe in human history'") and "city size" ("Humans might go the way of the trilobites"). She does concede that the latter catastrophe might strike "every 10 million to 30 million years," a time frame beyond human comprehension, but by raising the prospect, and describing its effect in contemporary terms — such "an asteroid landing in the Gulf of Mexico would cause floods in Kansas City," and rather more seriously, "entire continents" would "burst into flame" – the author emplaces it in the reader's head as a possibility to be considered and certainly feared when in fact it is immeasurably improbable.

Her final paragraph wraps alarmism and political advocacy into a potent package. "Killer asteroids and comets are out there," Begley writes, the homicidal adjective intimidating the reader toward her conclusion. "And someday, one will be on a collision course with Earth." (Probably not for millions of years.) "Of all the species that ever crawled, walked, flew or swam on Earth, an estimated two thirds became extinct because of an impact from space. Mankind may yet meet that fate, too. But we're the only species that can even contemplate it and, just maybe, do something to prevent it."[98]

Where do we send our check?

On a higher journalistic plane, the neoliberal science writer Gregg Easterbrook penned a lengthy story in the June 2008 issue of *The Atlantic Monthly* called, without a hint of understatement, "The Sky is Falling."

These stories typically follow a formula, and that formula, whether by design or not, scares the daylights out of readers right off the bat, so that the proffered "solutions" seem not flighty or ludicrously expensive but wise investments in the future of mankind. Take this opening scenario from Leonard David, writing in *Aerospace America*, which has never been mistaken for the *National Enquirer* or *Weird Tales*: A "huge rock billion of years old careens toward Earth." It hits land, and its "blast sparks monumental forest fires." Rock and debris are "flung high into the atmosphere" and are carried aloft by winds, so that the whole world is engulfed in acid rain and nitric-oxide smog. "When sunlight cannot penetrate this smog, darkness rules the day."[99] The day turns into years. The globe cools. Photosynthesis falters. The food chain breaks, and those human beings and animals not killed by the impact die off.

Who could possibly stand against doing anything and everything to avert such a catastrophe?

Again, this is not to say that a collision of Tunguska-like magnitude is impossible within our lifetimes or those of our children and grandchildren. *The Economist*, in praising the "search for dangerous asteroids," informed its readers in late 2008 that "experts in the field" figure that our chances of being slammed into by a Tunguska-type body within the coming century is one in ten, an overestimate by far, though only a fool would say that such a thing could never happen.[100] But there are reasonable worries and reasonable precautions and then there are "sky is falling" fables whose purpose is not so much moral instruction as it is the liberation of taxpayers' money from the Treasury to the private contractors and public agencies whose coffers bulge at the fable's climax.

The impact lobby was dismayed when in March 2007 NASA released a report to Congress titled "Near-Earth Object Survey and Deflection: An Analysis of Alternatives." This had been commissioned by the 2005 NASA Authorization Act, dubbed the "George E. Brown, Jr. Near-Earth Object Survey Act," after the California congressman who had served NASA well as a powerful advocate of space spending.

In the report, NASA's Office of Program Analysis and Evaluation recommended continuing the search for potentially hazardous NEOs, though with a modified goal of detecting and tracking by 2020 90 percent of those with diameters over 140 meters whose orbits pass within 0.05 AU (astronomical units), which is about 7.5 million kilometers or 4.65 million miles, of the Earth's orbit. This was seen as a retreat from new frontiers of science spending. The impact lobby wanted to track *all* NEOs of at least 140 meters, not just those that passed within the potentially hazardous threshold of 0.05 AU. This was an act of self-renunciation — in a very limited way — for the space agency. NASA, it seems, has other budgetary fish to fry. Although it expressed support for continuing Spaceguard, it politely declined any new NEO missions since "due to current budget constraints, NASA cannot initiate a new program at this time."

The NASA report was disdainful of the B612 Foundation's preferred method of asteroid deflection, saying that "'Slow push' mitigation techniques are the most expensive, have the lowest level of technical readiness, and their ability to both travel to and divert a threatening NEO would be limited unless mission durations of many years to decades are possible."[101] Rusty Schweickart took issue with the report's critical stance in testimony of October 11, 2007, before the Space and Aeronautics Subcommittee of the House Committee on Science and Technology, though he ended with an exaggerated exhortation: "We have the ability to make ourselves safe from cosmic extinction" — a statement of such supreme confidence as to make the rest of us knock extra hard on wood.[102]

The Spaceguard Survey folks at NASA calculate the frequency of a Tunguska-style collision at one every 300 years. Ronald Bailey of *Reason* magazine further explains that, given that much of the Earth is water or uninhabited (or nearly so) land, "another Tunguska is apt to hit a large urban area about once every 100,000 years." (Bailey, who favors spending more money on protection against near-Earth objects, does concede that "any new Planetary Defense Agency will suffer all the problems that afflict government bureaucracies, especially the innate drive to seek more money and power by exaggerating risks.")[103]

In any event, recall that the death toll from the Tunguska explosion was zero, which was also the death toll, as far as we know, from meteorite, asteroid, and comet impacts upon the Earth in the 20th century. It will almost certainly be the death toll in the 21st century, and the twenty-second as well. This is the unavoidable fact into which the asteroid lobby always runs.

The alarmist Duncan Steel appends to the title of his championing of Project Spaceguard the question, "Will Humankind Go the Way of the Dinosaurs?"

Not, apparently, if we fund asteroid-detection programs. Steel is somewhat more assertive of certainties than the record would warrant. "It is an unfortunate fact that the Earth is struck by a massive asteroid or comet from time to time, wreaking havoc on its inhabitants," he claims, though no such body has ever wreaked havoc on a human inhabitant. "[B]izarre as it may seem," he continues, "it is a fact [Mr. Steel is big on "facts" that others might dispute] that it is more likely that your gravestone will say that you died due to an asteroid impact than that you died in a jetliner crash."[104] Setting aside the minor point that fatalities in a collision with a massive asteroid are unlikely to have individual graves and gravestones, Steel reaches this actuarial conclusion by greatly reducing the frequency of catastrophic collisions. He figures that every 100,000 years, a quarter of the Earth's human population dies in such an event. Never mind that such an event has never occurred in the history of humankind, or that most estimate the infrequency of such an event at once every 50 million years. Duncan Steel says that at birth, you have a one in 5,000 chance of perishing due to a killer asteroid or comet, which is fifteen times greater than his calculation of one's chances of dying in an airplane crash. For heaven's sake, we use de-icers and flotation cushions, don't we? Shouldn't we employ their analogues to protect against doomsday? Or as Steel writes, "Governments quite rightly spend many millions of dollars ensuring that jetliners are properly maintained and operated, and that bombs or weapons are not smuggled aboard.... [W]hat should we be spending on making sure that an asteroid or comet does not catch us unawares?"[105]

There is hope, says Steel. For "Unlike the dinosaurs, we are smart enough to spot our potential nemesis coming, and then to shove it out of the way."[106]

Steel is among the most ardent of all planetary defenders, and his enthusiasm for the cause seems to magnify, with Palomar-like intensity, the actual dangers we face. At the end of this essay in *The Irish Astronomical Journal*, he comes clean, so to speak, revealing the awful imminence of the threat. A "minority of researchers on near-Earth objects," he writes, including himself in that minority, "believe that the actual hazard to civilization is much higher" than usually assumed "due to the arrival of a giant comet in a cis-jovian orbit about 20,000 years ago, the products of its subsequent disintegration having a significant influence upon the terrestrial planets, including the affairs of humans." As a result, the real chances of our civilization facing "extraterrestrial incursions" — a strange choice of words — in our lifetimes is one in ten or twenty — "or even higher."

This is an extraordinary admission, and one that is so much at odds with what we know about asteroids and comets that it places Steel almost in a galaxy of his own. Of course challenges to conventional wisdom are all to the good, but the consequences of Steel's heterodoxy — his belief that the end of the world as we know it is quite possibly imminent — lend themselves to hyperalarmism, to put it mildly. He is very enthusiastic about Spaceguard, and believes it an "abrogation of our responsibilities to our fellow humans" to not fully fund (and more) such programs.[107]

"Planetary defense" is becoming a buzzword in this burgeoning field. It starts with telescopic surveys of the sky, as near-earth asteroids of greater than one kilometer in diameter and the orbits of comets are tracked and mapped with the use of CCDs (charge-coupled devices), which have largely replaced photography as an astronomer's tool. Forewarned is forearmed, say the planetary defenders. Few object to the compilation of NEOs and their orbits and, in the event that one of these bodies is hurtling toward a once in a million-plus year collision with Earth, devising a plan for deflection of the interloper.

Among those doing surveillance work in near-earth asteroids is the Panoramic Survey Telescope & Rapid Response System, or Pan-STARRS, an Air Force-funded University of Hawaii undertaking that utilizes small mirrors and large digital cameras to "discover and characterize Earth-approaching objects, both asteroids & comets, that might pose a danger to our planet." These cameras, each of which has 1.4 billion pixels (compared to five million pixels for the run-of-the-mill digital camera), will permit the Hawaii researchers to sweep larger portions of the sky than is possible with conventional instruments.[108] The estimate is that within a decade, Pan-STARRS will have ferreted out most of the asteroids and comets that are within the seeing range of Hawaii. A worthy project, no doubt. But projects, however worthy, need funding, and if Buck Rogers is necessary for the funding of space travel, the Armageddon-bearing celestial ball of rock is the ticket when it comes to sky surveys.

"There is a significant potential that the Earth will be hit by a killer asteroid," Dr. Rolf Kudritzki, director of the Institute for Astronomy at the University of Hawaii, told the PBS viewing audience during *400 Years of the Telescope*, a much-promoted feature in 2009. "You really have to worry about this," he continued, using the factoid that the probability that a person will die due to an asteroid collision with Earth is comparable to the chance that he will die in an airplane crash. "With Pan-STARRS, we will be able to detect such killer asteroids thirty years before impact. That will give mankind an opportunity to do something about it and to avoid impact by a killer asteroid."

Dr. Kudritzki's Institute for Astronomy benefits substantially from a heightened fear of killer asteroids. Pan-STARRS will consist of four telescopes, the first of which went into service in December 2008 and the last of which is scheduled to see first light by 2012. The $100 million project is being subsidized by the U.S. Air Force, which has firmly established itself at the top of the preventing-the-end-of-the-world bureaucratic totem pole.

The ever-candid Carl Sagan gave us a glimpse into his motivation for counseling vigilance in the matter of approaching comets and asteroids. Besides protecting Earth from Armageddon, which seems about as uncontroversial a cause as one could find, Sagan believed — hoped — that "the small near-Earth worlds provide a new and potent motivation to create effective transnational institutions and to unify the human species."[109] The United Nations has had sixty years to do the job and national rivalries have yet to melt into universal brotherhood, so perhaps it's time to give NASA a chance? (This was very much in the spirit of the musing of Sagan's bête noire, President Ronald Reagan, who in talking about his "Star Wars" defense proposal speculated about the incentives for peaceful cooperation between the United States and the Soviet Union which a threatened alien attack would bring.)

"The asteroid hazard forces our hand," wrote Sagan. To meet this hazard, "we must establish a formidable human presence throughout the inner Solar System," and to "do so safely we must make changes in our political and international systems."[110]

As his fellow liberal internationalist space advocate Daniel Deudney wrote in support of a U.S–Soviet asteroid exploration project, "it could be critical to human survival."[111]

It takes either the most corroded cynic or the most luminous idealist to try to turn the destruction of the world into policy reforms. The reader can decide for herself about Sagan and the Air Force, but they were not alone in sensing opportunity. The Department of Homeland Security, the mega-bureaucracy born in the ashes of 9/11, is keeping a turf-building eye out for alleged threats to the United States that go well beyond the bounds of terrorism. Certainly fanatical jihadists smuggling dirty bombs into Manhattan are frightening, but they are nowhere near as compelling as the rogue doomsday asteroid or comet hurtling through the immense void of space right at a helpless Earth.

The DHS was quick to grasp the budget-swelling potential of killer comets. Established in 2002 in the panic following the 9/11 attacks, the department was but a toddler in bureaucratic years when it developed the "National Response Plan" called for by Presidential Directive/HSPD-5 of February 2003, in which President Bush asked this new and unwieldy bureaucracy to oversee development of "a single comprehensive approach to domestic incident management."[112] ("Domestic incident management"? Aren't we prim?) These "incidents" were said to include "terrorist attacks, major disasters, and other emergencies," those "other emergencies" including rocks falling from the sky.

This was a great boon for this new department, which would be coordinating a response that stretched from NASA and the military to FEMA, the National Institutes of Health, and the whole range of military and health-related agencies. It was a prime opportunity for the DHS to expand its turf. Fanatical Muslims will eventually lose their fanaticism, after all, or at least their capacity to frighten, but rogue comets and asteroids: these have staying power!

As is usually the case with government agencies, the language is curious. Managing "domestic incidents" sounds like intervening in husband–wife spats, not evacuating 300 million American casualties of Target Earth. The plan envisions a "single, comprehensive approach" to such "incidents," consistent with the one-size-fits-all philosophy of the erstwhile Eastern Bloc. It also draws upon such unsavory documents as the "Mass Migration Emergency Plan," which envisions the transfer of large populations in the event of emergencies.

Now, it makes sense for the appropriate agencies to make plans for the evacuation of cities in the event of a levee breaking or a power plant disaster. These "incidents" are plausible, or at least thinkable. Better safe than sorry. A collision with an asteroid or comet, on the other hand, is so highly implausible, so exceedingly unlikely, that "planning" for it is a potent blend of the useless and the expensive.

Boosting the DHS's National Response Plan was Evan R. Seamone, writing in 2004 in the *Georgetown International Environmental Law Review*. Seamone bemoaned that "Current legal and policy efforts to enable adequate defense against potential asteroid or comet collisions with the earth are insufficient because they are indirectly premised upon theories that require verification of a clear and imminent threat before governmental agencies can act." In other words, the antediluvian theories that underlay our system and the systems of most other governments of the world require that there be an actual *threat* before the state is mobilized to meet that threat. Obviously the philosophers who spin such theories never saw *Armageddon* on DVD. As an alternative for the 21st century, Seamone proposed a "precautionary principle" as the cornerstone of a governmental asteroid defense program. This tenet — which might also be known as the fling-open-the-doors-to-the-Treasury principle — "requires governments to take action to prevent harm even when it is uncertain if, when, or where the harm will occur."

As the hook for his article, Seamone mentions the killer asteroid scare of early 2004, as the unpoetically named 2004 AS1 was briefly and inaccurately said to be hurtling toward earth. It was not, though Seamone says that "the public came very close to knowing the horror of an impending asteroid disaster first-hand" in January 2004. This nonthreat, which had been wrongly reported as a threat, ought to have jolted the nations of the world, and in particular the United States, into action, thought Seamone. He urged the government to mobilize for "planetary protection," with a major role for the Department of Homeland Security.

Seamone called for the development of a "single Planetary Defense Plan" to ensure that the response of the U.S. government to the possible end of the world was not fragmented. From coordinating telescopic surveillance of potential world-enders to evacuating those luckless enough to be in the path of the tsunamis and earthquakes that the collision would cause, Washington needs to enact "greater, more meaningful, regulation of space harm," he declared. ("Space harm" is a fine, almost absurdly understated euphemism for "killer asteroids.") For instance, NASA lacks the power "to deploy nuclear devices" — an oversight that, if corrected, would empower the space agency beyond the wildest dreams of its supporters.[113]

That would have been martial music to the ears of Edward Teller, whose name seems to recur whenever the subject is federally subsidized science. Hydrogen bomb developer Teller took an interest in asteroid/comet defense late in his life. Teller proposed "the development of a bomb 10,000 times more powerful than anything in today's arsenal," a course of action that would be extraordinarily expensive, of questionable usefulness, and which would guarantee to its owner a kind of absolute veto over world affairs — at least if its owner had a suicidal streak.[114]

In a paper coauthored with NASA's David Morrison, Teller let his imagination run wild. After all, from *Sputnik* to the necessity of imposing the metric system on Americans, Teller's name has often been found wherever hysteria in the service of government expansion is sold.

In "The Impact Hazard: Issues for the Future," Teller (and Morrison) kicked things off by stating that "impacts from asteroids and comets constitute a credible and significant hazard to life and property." Well — who says? The authors say so, tautologically.

They admit that "there are no credible historical records of human casualties from impacts in the past millennium," yet they insist that the expenditure of "public funds for a program of NEO discovery and research is justified" given what the federal government spends on other science and risk reduction programs.[115]

Complications present themselves. What if, for example, a tyrant got hold of such a doomsday device? Well, no plan is perfect. That's for the next generation to worry about.

The entire disaster management industry — for instance, the supremely competent FEMA bureaucrats who responded with such alacrity to Hurricane Katrina — smells opportunity in the Armageddon from the skies scenario. Writing in *Space Policy*, that same David Morrison of NASA's Ames Research Center, Victoria Garshnek of Global Human Futures Research Associates, and Frederick M. Burkle, Jr. of the Department of Emergency Medicine at the University of Hawaii insist that the "asteroid/comet impact hazard is a realistic threat to the human population." Proceeding from this highly questionable opening sentence, the authors discuss "the NEO impact hazard as a public health issue."[116] They concede that public health agencies can do little about Earth-demolishing monster rocks, and the meteorites that rain like dust on the Earth every hour of every day pose no hazard. The problem, as they see it, lies with the intermediate-size bodies with diameters between a few dozen meters to hundreds of kilometers.

The smallest Earth-striking body that would have public health consequences, they say, would be about 10 meters in diameter. This would cause localized damage, blast injuries, and possibly fires. The authors caution that "[s]hort-term mass hysteria control" may be necessary, especially if residents of the affected area have seen *Armageddon* and *Deep Impact*.[117] Proceeding up the damage scale, they find more and more for governments and the military to do, whether in managing (rationing) medical and food resources, evacuating populations unlucky enough to live near the impact site, and doing all sorts of things on a global scale that governments currently find difficult enough to do within their own boundaries.

The largest conceivable impact is from a body 5 or more kilometers in diameter. It would bring, they say, the end of the world as we know it, as the fires and flooding and blocking of sunlight would plunge the Earth into blackness and death and result in "mass mortality, mass extinctions."[118] One cannot read of this hell without a shiver, and without thinking that perhaps we ought to give the defenders of Earth whatever they want, damn the cost. That the authors project such an event to occur once every 10–30 million years gets sort of lost in the sheer terror that the description evokes.

And yet no matter how often Garshnek, Morrison, and Burkle insist that "the NEO threat can be considered a public health threat," they are forced to concede that an "impact threat may not be immediately apparent since a moderate-to-major impact has not occurred within human history or memory."[119]

Well, yes, that does pose an obstacle to taking this threat seriously. The authors may insist that doing nothing is a selfish attitude reeking of "let it be a problem for future generations to deal with," but there are many, many,

many "future generations" packed into 30 million years. Skeptics of an enhanced Earth protection plan, they charge, are failing to ask themselves if "human civilization [is] worth saving. If everything we have been a part of in our lifetime and historically evolved from [is] worth preserving?"[120]

If you would vote against, say, doubling the NEO detection budget, or adding asteroid-impact evacuation planning to the portfolio of the Department of Homeland Security, you are indifferent to the entirety of human history, human culture, human creation. You are a first-class boor.

This chapter began with a killer-asteroid scare and shall end the same way. The story of 2004 ASI, the Earth-killer that wasn't, illustrates again the ways in which hysteria can be used to further a political or budgetary cause. The asteroid was discovered by the LINEAR sky survey. Its coordinates were sent out by Brian Marsden's Minor Planet Center of the International Astronomical Union, based at the Smithsonian Astrophysical Observatory in Cambridge, Massachusetts, the clearinghouse for information on subplanetary bodies. A researcher at the Jet Propulsion Laboratory, Steven Chesley, calculated that 2004 ASI "had a 25 percent chance of striking the Earth's Northern Hemisphere in a few days," according to the British Broadcasting Company (BBC). Given that it was first believed to be 30 meters in diameter (and later found to be much larger), a collision would have caused significant death and destruction on the home planet.

Asteroid researchers Clark Chapman and David Morrison, alarmed by Chesley's doomsday calculation, "contemplated picking up the telephone to the White House." They were going to alert President Bush that the end times were closer than they appeared in his rearview mirror. Perhaps Chapman and Morrison had seen too many movies in which a scientist, having seen an ominous spot growing larger by the second in his secluded mountaintop observatory, picks up the telephone, rings the operator, says, "Get me the President of the United States," and then the film cuts to the President informing his Cabinet that he has to tell them about a matter of "grave importance."

In any event, Chapman and Morrison held off on that call, and it was a good thing. The killer-asteroid 2004 ASI missed the Earth by 12 million kilometers, or more than 30 times the distance between Earth and moon. It wasn't even a close shave. As the BBC reported, many astronomers harshly criticized Chapman and Morrison for almost ringing "a false alarm [which] could have brought ridicule on their profession."

For instance, Benny Peiser of Liverpool's John Moores University told the BBC, "They completely misread the situation. There was plenty of time to get other observers on the job."

Brian Marsden of the Minor Planet Center, who had been castigated for his own overreaction in the case of XF 11, said, "They would have jumped the gun before we knew much about the object. I find it incredible that such action was contemplated on the basis of just four observations. That is just enough to yield a sensible orbit. There was no need to panic as it was obvious that the situation would have been resolved, one way or another, in another hour or two."[121]

So an embarrassing act of Chicken Little-ism was averted. The sky was not falling, and the alarmists took a little public criticism for what would have been a grandiose indiscretion.

The following year, Congress authorized more near-earth object detection efforts, though as Gregg Easterbrook writes, the price tag that NASA put on such a mission — $1 billion — was too steep, given the agency's bigger-hardware priorities such as a moon base. (Easterbrook nails the real rationale behind the moon base, which, on its surface, is a distraction from the vaunted mission to Mars: "For NASA, a decades-long project to build a moon base would ensure a continuing flow of money to its favorite contractors and to the congressional districts where manned-space-program centers are located."[122] Pork must be served.)

Chapman and Morrison have pondered NEOs for many years now, and they admit the inherent ambiguities. Clark Chapman concedes that "there is deep disagreement over whether we should also protect against the impacts that happen every decade or so, like Tunguska" — though the last Tunguska happened not a decade but a century-plus ago. "Even these small events can kill people, but they are a thousand times less likely to do so than are quakes, floods and the other things that kill people all the time."[123]

David Morrison says, "It's truly an apocalyptic vision that you have here," but he concedes that "there are very human reactions as to whether this one-in-a-million-per-year risk [which may be an exaggerated number itself] is worth worrying about or not."[124] Clark Chapman adds that "such once-in-100 million year events are so rare that, despite their apocalyptic horror, they need be of no concern to public officials."[125] (Note the sharp difference in estimates of the chances of a civilization-ending collision.)

If a one-in-a-million — or 65 million, or one trillion — year doomsday comet suddenly raced in from the Oort Cloud, there is simply no defense known or even contemplated against it. We would be out of luck. Yet as a team of researchers wrote in *Reviews of Geophysics*, asteroid and comet collisions "are so infrequent that they are normally disregarded on the timescale of human evolution."[126] Prudence dictates that we not entirely ignore the incredibly remote possibility that such a collision could happen at any time during the next 40 million years, but that same prudence should keep us from panic, and prevent us from public expenditures that cannot be justified by any wisdom this side of sheer Hollywood-sized hysteria.

Even without a rogue asteroid banging into the Earth, life as we know it will be impossible on the planet in a billion or more years, when the Sun swells 250 times its current size, into a "red giant" star that will swallow our home planet.[127] If you wish to worry about that, fine. Same for those who stay up nights pulling out their hair over the prospect of an Armageddon asteroid. But the rest of us — at least those of us who do not make our living in the NEO detection field — have quite enough else to worry about, including a swelling budget deficit whose size may soon dwarf the rockiest chunks in the Asteroid Belt.

Notes

1. Sagan, *Pale Blue Dot*, p. 323.
2. Clark R. Chapman, "The Asteroid/Comet Impact Hazard: Homo Sapiens as Dinosaur?" in *Prediction: Science, Decision Making, and the Future of Nature*, edited by Daniel Sarewitz, Roger A. Pielke Jr., and Rayford Byerly Jr. (Covelo, CA: Island Press, 2000), p. 111.
3. Martin Gardner, "Near-Earth Objects: Monsters of Doom?" *Skeptical Inquirer* (July/August 1998), p. 16.
4. Chapman, "The Asteroid/Comet Impact Hazard: Homo Sapiens as Dinosaur?" p. 110.
5. Bonnie Bilyeu Gordon, "That Asteroid Caper," *Astronomy* (July 1998), p. 6.
6. Chapman, "The Asteroid/Comet Impact Hazard: Homo Sapiens as Dinosaur?" pp. 109, 113.
7. Gordon, "That Asteroid Caper," p. 6.
8. Chapman, "The Asteroid/Comet Impact Hazard: Homo Sapiens as Dinosaur?" pp. 113, 126, 121.
9. Philip Plait, *Death from the Skies!* (New York: Viking, 2008), pp. 4, 9.
10. Robert S. Boyd, "Scientists Seek Ways to Ward Off Killer Asteroids," McClatchy Newspapers, December 17, 2008.
11. Luis W. Alvarez, Walter Alvarez, Frank Asaro, and Helen V. Michel, "Extraterrestrial Cause for the Cretaceous-Tertiary Extinction," *Science*, Vol. 208, No. 4448 (June 6, 1980), pp. 1095, 1106.
12. Ibid., p. 1095.
13. Plait, *Death from the Skies!*, p. 16.
14. Alvarez et al., "Extraterrestrial Cause for the Cretaceous-Tertiary Extinction," p. 1105.

15. Peter M. Sheehan and Dale A. Russell, "Faunal Change Following the Cretaceous-Tertiary Impact: Using Paleontological Data to Assess the Hazards of Impacts," in *Hazards Due to Comets & Asteroids*, edited by Tom Gehrels (Tucson: University of Arizona Press, 1994), p. 881.

16. Dennis V. Kent, "Asteroid Extinction Hypothesis," *Science*, Vol. 211, No. 4483 (February 13, 1981), p. 649.

17. Ibid., p. 650.

18. Richard A. Kerr, "Periodic Extinctions and Impacts Challenged," *Science*, Vol. 227, No. 4693 (March 22, 1985), pp. 1451–53.

19. Alvarez et al., "Extraterrestrial Cause for the Cretaceous-Tertiary Extinction," p. 1107.

20. Edgar Allan Poe, "The Conversation of Eiros and Charmion" in *Tales of Mystery and Imagination* (New York: Brentano's, 1928), p. 163.

21. Ibid., p. 166.

22. H.G. Wells, *In the Days of the Comet* (Lincoln: University of Nebraska Press, 2001/1906), pp. 162, 169.

23. Gardner, "Near-Earth Objects: Monsters of Doom?" p. 17.

24. Marc Davis, Piet Hut, and Richard A. Muller, "Extinction of Species by Periodic Comet Showers," *Nature*, Vol. 308 (April 19, 1984), p. 717.

25. Ibid., p. 715.

26. Duncan Steel, *Target Earth* (Pleasantville, NY: Reader's Digest, 2000), pp. 25, 28.

27. "Impact History," http://www.b612foundation.org.

28. Sagan, *Pale Blue Dot*, pp. 313–14.

29. Charles Q. Choi, "Could Earth Be Hit, Like Jupiter Just Was?" SPACE.com, July 28, 2009.

30. James V. Scotti, "On Comets," *The Scientific American Book of the Cosmos*, edited by David H. Levy (New York: St. Martin's, 2000), p. 185.

31. Sharon Begley, "The Science of Doom," *Newsweek* (November 23, 1992).

32. Gregg Easterbrook, "The Sky Is Falling," *The Atlantic* (June 2008), p. 76.

33. Sandra Blakeslee, "Ancient Crash, Epic Wave," *New York Times*, November 14, 2006.

34. Easterbrook, "The Sky Is Falling," pp. 76–80.

35. "Space Rock Gives Earth a Close Shave," AFP, March 3, 2009, http://www.breitbart.com.

36. Robert S. Boyd, "Scientists Seek Ways to Ward Off Killer Asteroids."

37. "B612 Foundation Statement Regarding NASA's Analysis of Asteroid 99942 Apophis Impact Potential," http//www.b612foundation.org.

38. "Near Earth Object Program," http//neo.jpl.nasa.gov/torino_scale.html.

39. Quoted in Easterbrook, "The Sky Is Falling", p. 80.

40. Ibid., pp. 82–83.

41. http://www.b612foundation.org.

42. Easterbrook, "The Sky Is Falling," p. 84.

43. http://www.b612foundation.org.

44. Gregory H. Canavan, Johndale C. Solem, and John D.G. Rather, "Near-Earth Object Interception Workshop," in *Hazards Due to Comets & Asteroids*, p. 93.

45. Thomas J. Ahrens and Alan W. Harris, "Deflection and Fragmentation of Near-Earth Asteroids," *Nature*, Vol. 360 (December 3, 1992), pp. 430–32.

46. Clark R. Chapman, Daniel D. Durda, and Robert E. Gold, "The Comet/Asteroid Impact Hazard: A Systems Approach," (San Antonio, TX: Southwest Research Institute, February 24, 2001), p. 13.

47. Ahrens and Harris, "Deflection and Fragmentation of Near-Earth Asteroids," p. 433.

48. A.V. Bushman, A.M. Vickery, V.E. Fortov, B.P. Krukov, I.V. Lomonosov, S.A. Medin, A.L. Ni, A.V. Shutov, and O.Yu. Vorobiev, "Computer Simulation of Hypervelocity Impact and Asteroid Explosion," in *Hazards Due to Comets & Asteroids*, p. 718.

49. Ahrens and Harris, "Deflection and Fragmentation of Near-Earth Asteroids," pp. 430, 433.

50. Quoted in Leonard David, "Assessing the Threat from Comets and Asteroids," *Aerospace America* (August 1996).

51. V.A. Simonenko, V.N. Nogin, D.V. Petrov, O.N. Shubin, and Johndale C. Solem, "Defending the Earth Against Impacts from Large Comets and Asteroids," in *Hazards Due to Comets & Asteroids*, p. 930.

52. Ibid., p. 951.

53. Joseph G. Gurley, William J. Dixon, and Hans F. Meissinger, "Vehicle Systems for Missions to Protect the Earth Against NEO Impacts," in *Hazards Due to Comets & Asteroids*, p. 1035.

54. Alan W. Harris, Gregory H. Canavan, Carl Sagan, and Steven J. Ostro, "The Deflection Dilemma: Use Versus Misuse of Technologies for Avoiding Interplanetary Collision Hazards," in *Hazards Due to Comets & Asteroids*, p. 1145.

55. P.R. Weissman, "The Comet and Asteroid Impact Hazard in Perspective," in *Hazards Due to Comets & Asteroids*, p. 1191.

56. Ibid., p. 1196.

57. Robert L. Park, Lori B. Garver, and Terry Dawson, "The Lesson of Grand Forks: Can a Defense Against Asteroids be Sustained?" in *Hazards Due to Comets & Asteroids*, pp. 1226–27.

58. Chapman, Durda, and Gold, "The Comet/Asteroid Impact Hazard: A Systems Approach," p. 16.

59. Owen B. Toon, Kevin Zahnle, David Morrison, Richard P. Turco, and Curt Covey, "Environmental Perturbations Caused by the Impacts of Asteroids and Comets," *Reviews of Geophysics*, Vol. 35, No. 1 (February 1997), p. 75.

60. Ibid., pp. 61, 41.

61. Sheehan and Russell, "Faunal Change Following the Cretaceous-Tertiary Impact: Using Paleontological Data to Assess the Hazards of Impacts," in *Hazards Due to Comets & Asteroids*, p. 890.

62. Toon, Zahnle, et al., "Environmental Perturbations Caused by the Impacts of Asteroids and Comets," p. 44.

63. Neal S. Young, John P.A. Ioannidis, and Omar Al-Ubaydli, "Why Current Publication Practices May Distort Science," *PLoS Medicine*, Vol. 5, Issue 10 (June 2008), p. 1418.

64. "Publish and Be Wrong," *The Economist* (October 11, 2008), p. 109.

65. Young, Ioannidis, and Al-Ubaydli, "Why Current Publication Practices May Distort Science," pp. 1418–21.
66. Michael B. Gerrard, "Risks of Hazardous Waste Sites versus Asteroid and Comet Impacts: Accounting for the Discrepancies in U.S. Resource Allocation," *Risk Analysis*, Vol. 20, No. 6 (2000): 899.
67. Ibid., pp. 901–902.
68. Ibid., p. 903.
69. S. Nozette, L. Pleasance, D. Barnhart, and D. Dunham, "DoD Technologies and Missions of Relevance to Asteroid and Comet Exploration," in *Hazards Due to Comets & Asteroids*, p. 671.
70. Quoted in Chris Ferenzi, "Mars or Bust," February 4, 2004, http://redcolony.com.
71. David, "Assessing the Threat from Comets and Asteroids."
72. Steel, *Target Earth*, p. 155.
73. Ibid., pp. 127, 135.
74. Ibid., p. 71.
75. Ibid., pp. 6–7.
76. Ibid., pp. 8.
77. Ibid., pp. 128, 130.
78. Ibid., pp. 20, 152–53.
79. Choi, "Could Earth Be Hit, Like Jupiter Just Was?"
80. Chapman, Durda, and Gold, "The Comet/Asteroid Impact Hazard: A Systems Approach," pp. 1–3.
81. Ibid., pp. 10–11.
82. Clark R. Chapman, "How a Near-Earth Object Impact Might Affect Society," Global Science Forum (January 2003): n.p.
83. David Morrison, Clark R. Chapman, and Paul Slovic, "The Impact Hazard," in *Hazards Due to Comets & Asteroids*, p. 81.
84. Ibid., pp. 82–84.
85. Ibid., p. 87.
86. Weissman, "The Impact Hazard in Perspective," p. 1203.
87. Ibid., pp. 1208–1209.
88. William K. Hartmann and Andrei Sokolov, "Evaluating Space Resources in the Context of Earth Impact Hazards: Asteroid Threat or Asteroid Opportunity?" in *Hazards Due to Comets & Asteroids*, p. 1213.
89. Ibid., pp. 1218, 1223.
90. Andrew F. Cheng, J. Veverka, C. Pilcher, and Robert W. Farquhar, "Missions to Near-Earth Objects," in *Hazards Due to Comets & Asteroids*, p. 651.
91. T.D. Jones, D.B. Eppler, D.R. Davis, A.L. Friedlander, J. McAdams, and S. Krikalev, "Human Exploration of Near-Earth Asteroids," in *Hazards Due to Comets & Asteroids*, p. 687.
92. John Mueller, *Overblown: How Politicians and the Terrorism Industry Inflate National Security Threats, and Why We Believe Them* (New York: Free Press, 2006), p. 25.
93. Ibid., p. 37.
94. Sagan, *Pale Blue Dot*, p. 306.

95. McDougall, "Technocracy and Statecraft in the Space Age — Toward the History of a Saltation," p. 1035.
96. Begley, "The Science of Doom."
97. Sally Stephens, "Cosmic Collisions," *Astronomical Society of the Pacific* (Spring 1993), http://www.astrosociety.org.
98. Begley, "The Science of Doom."
99. David, "Assessing the Threat from Comets and Asteroids."
100. "Watching and Waiting," *The Economist* (December 4, 2008).
101. "Near-Earth Object Survey and Deflection: An Analysis of Alternatives," March 2007, http://neo.jpl.nasa.gov.
102. "Testimony of Russell L. Schweickart Before the Space and Aeronautics Subcommittee of the House Committee on Science and Technology," October 11, 2007, p. 13, http://www.b612foundation.org.
103. Ronald Bailey, "Earth Killers from Outer Space," *Reason*, January 26, 2005, http://www.reason.com.
104. D. Steel, "Project Spaceguard: Will Humankind Go the Way of the Dinosaurs?" *Irish Astronomical Journal*, Vol. 24, No. 1 (1997), p. 19.
105. Ibid., p. 26.
106. Ibid., p. 19.
107. Ibid., p. 29.
108. For more on Pan-STARRS, see http://pan-starrs.ifa.hawaii.edu.
109. Sagan, *Pale Blue Dot*, p. 263.
110. Ibid., p. 264.
111. Deudney, "Forging Missiles into Spaceships," p. 295.
112. "Initial National Response Plan," U.S. Department of Homeland Security, September 30, 2003, p. 2.
113. Evan R. Seamone, "The Precautionary Principle as the Law of Planetary Defense: Achieving the Mandate to Defend the Earth Against Asteroid and Comet Impacts While There Is Still Time," *Georgetown International Environmental Law Review* (Fall 2004), n.p.
114. Begley, "The Science of Doom."
115. David Morrison and Edward Teller, "The Impact Hazard: Issues for the Future," in *Hazards Due to Comets & Asteroids*, pp. 1135, 1137.
116. Victoria Garshnek, David Morrison, and Frederick M. Burkle Jr., "The Mitigation, Management, and Survivability of Asteroid/Comet Impact with Earth," *Space Policy*, Vol. 16 (2000), pp. 213–14.
117. Ibid., p. 216.
118. Ibid., p. 217.
119. Ibid., p. 215.
120. Ibid., p. 221.
121. David Whitehouse, "Earth Almost Put on Impact Alert," BBC News, March 24, 2004, http://newsvote.bbc.co.uk.
122. Easterbrook, "The Sky Is Falling," p. 82.
123. Begley, "The Science of Doom."
124. David, "Assessing the Threat from Comets and Asteroids."

125. Chapman, "How a Near-Earth Object Impact Might Affect Society."

126. Toon et al., "Environmental Perturbations Caused by the Impacts of Asteroids and Comets," p. 41.

127. Jeremy Hsu, "Scientists Speculate on Earth's Last Day," Space.com, August 3, 2009, http://news.aol.com.

Chapter 7

Conclusion: The Only Way to Keep the Sky from Falling Is…

Killer asteroids, trips to Mars, Russian satellites arcing across the sky: we've come a long way from the debates over John Quincy Adams's "light-houses of the sky" — or even Vannevar Bush's National Science Foundation. Big Science has learned well how to obtain funding for even outlandish-sounding projects.

But there also exists a counter-tradition of constitutionalist skepticism that views both National Defense Education Acts (NDEAs) and Near-Earth Asteroids (NEAs) — not to mention the National Education Association — with a critical eye. That tradition has become a distinctly minority affair, but in the coming age of missions to Mars and planetary defense schemes, we may need it more than ever. The time has come to hear from the Frank Jewetts again.

This tradition, it is important to emphasize, is not obscurantist, or anti-science, or anti intellectual and does not disparage intelligence or belittle the scientific method. It does not substitute its own set of preordained conclusions for those of the politicized and politically savvy scientists. It does not say that rogue comets should not be monitored, or Mars should not be studied, or physics is not an important subject for high-school students, or that our men and women of science should not continue to push on to new frontiers, cross new boundaries of knowledge. *It is not in any way, shape, or form anti-science.*

Rather, this tradition is grounded in a healthy skepticism of state power. It does not view government as savior or as a crusading force for enlightenment. It does not credit government with discovery or wisdom or discernment. It views government as a mechanism through which wealth can be redistributed to the most politically potent actors. For many decades in the United States, our Constitution and the attitudes that shaped it prevented

J.T. Bennett, *The Doomsday Lobby: Hype and Panic from Sputniks, Martians, and Marauding Meteors*, DOI 10.1007/978-1-4419-6685-8_7,
© Springer Science+Business Media, LLC 2010

the transfer of great sums of wealth to scientific institutions and individuals. The creation of a scientific bureaucracy was achieved in fits and starts, but by the post-World War II era, with the birth of the National Science Foundation, the old constitutional restraints no longer held. The only real limits upon the growth of government science were fiscal. Science had its place at the table. The pie was before it. The only question was how big a slice might be cut. In arguing for a bigger piece, politicized scientists took to exaggeration, the conjuring of threats, the tying of scientific research to defense needs, and fierce lobbying. It is difficult to see how it could have been otherwise once the constitutional wall separating government from science had been so totally breached.

Piecemeal reforms of the current system are bound to fail. Greater oversight of NSF grants, tighter regulations on the use of grant monies, shifting emphases within science education: these will not even glancingly affect the central problem. If a threat (or deliverance from a threat) can be magnified and publicized enough, funding will flow like a Niagara Falls-force cataract. Convince enough people – or politicians – that the world is going to end, and the world's pocketbook is yours.

This phenomenon is hardly limited to the United States. Russia, from whose Soviet incarnation the advocates of government science drew so much inspiration in the 1940s, is undertaking a ludicrous asteroid diversion plan intended to keep Apophis (a.k.a. "the Continent Killer" and "Doomsday 2036") from striking the Earth. You will recall from the previous chapter that NASA had reckoned the odds of Apophis causing mischief at 44,000 to 1. In late 2009, as Anatoly Perminov of the Russian space agency Roscosmos announced that his bureaucracy had Apophis in its crosshairs, NASA recalculated those odds at 250,000 to one. Yet since Perminov held out the prospect of international cooperation to divert the asteroid, we can expect the killer asteroid community to treat Apophis with new respect. After all, it just might be made of gold.

We should not expect otherwise. For as long as there is loot to be had, the Chicken Littles will squawk. Scientists have been trying to pry open public treasuries for their own personal gain ever since public treasuries were amassed. The only real way to end the unseemly, sometimes dishonest, often counterproductive scramble for funds is to cease handing them out – to return the federal government to its constitutional role, which in the field of science means the granting of patents and not much else. Science must once again be the province of individual genius, cooperative research, philanthropy, universities, dedicated amateurs, profit-minded entrepreneurs – the whole gamut of persons and institutions whose efforts formed the foundation of American science.

In the United States, the lid upon the public treasury — or "lockbox," to use a certain former Vice President and global-warming promoter's phrase — remained closed for many a year. American scientists, funded in the marketplace, by foundations, by philanthropists, or working with little recompense beyond the satisfaction one receives from the pursuit of scientific and intellectual truths, achieved extraordinary things. In the field of astronomy, as we have seen, Americans were outpacing the rest of the world by the late 19th century, despite an almost total lack of federal support of the astronomical sciences. Geologists, chemists, physicists: American scientists took their place in the vanguard of intellectual advancement without the patronage of kings and queens, of empires and bureaucracies. No federal Department of Science was necessary to these achievements.

In the post-World War II era, the relationship between state and science shifted profoundly. Science was now both client and protector of the federal government. Washington subsidized laboratories, scientists, and classrooms; it undertook massive projects, yet its reach also extended into high schools and movie theaters. It was all that stood between us and Armageddon. Those who questioned this arrangement were patronizingly consigned to the 19th century — they were said to be men out of touch with new realities, with modern ways of doing things.

In fact, it was those dissidents who understood the dangers of concentrating too much power in the central government, and making scientists the clients of that government. Neither science nor scientists benefit from such an arrangement.

Potted science, driven by political agendas, increasingly drives federal funding. As this book was being written, a shocking scandal broke — shocking, that is, to those who are not aware of the at times seamy history of government-funded science. It is a prime example of the corruption that can seep in when scientists pursue the lucre of state subsidy.

The scandal is centered on that most mundane of subjects, the thing that everybody talks about but nobody does anything about: the weather. Except in this case, not only were government-funded scientists talking about it, they intended to *do* something. They intended to change it.

I refer, of course, to the debate over anthropogenic — that is, caused by human action — global warming, upon which a good deal of light, heat, and even darkness has been shed over the last decade. In this brief span of years, the theory of global warming has hardened into orthodoxy whose high priests permit no dissent. They could have taught Galileo's persecutors a thing or two.

The theory, most famously and effectively promoted by former Vice President Al Gore in his documentary *An Inconvenient Truth* (2006), holds that our planet's climate is changing, and that a significant cause of such change is man-made emissions of such greenhouse gases as carbon dioxide. If, in fact, such climate change is occurring, a range of policy alterations recommend themselves – most of them empowering the governments of the world, and the world's government, in the form of the United Nations – to take drastic action to sharply reduce carbon dioxide emissions. These actions might include unprecedented global taxes and the creation of new science bureaucracies whose funding would be so immense as to make it appear that million-dollar grants were falling from the sky like raindrops.

The facts of the case are well beyond the scope of this book. There are many fine (and not so fine) books on climate change. Suffice to say that our planet's temperature does seem to have been rising, although there is disagreement over the degree and significance of this rise, its duration, its causes, and its implications. There is also some evidence, though it is disputed, that this increase is coming to a halt – that global warming is *not* the current state of the world.

Perhaps climate change is man-made, or perhaps it is cyclical. For instance, climate scientists have long known of the globe-wide warming during the medieval period of about 1,000 years ago, when Greenland was not icy but truly green with vegetation and cropland. A mini-Ice Age is then said to have occurred from about 1500 to 1850, as the Earth's temperature cooled. Whether random or part of a natural cycle, these earlier changes were obviously unrelated to the industrial revolution, automobiles, or similar human activities.

Whatever the cause or causes, these are matters for serious scientific inquiry: for advancing and testing hypotheses, for gathering and considering data, for informed exchange and even argument.

Instead, the skeptics of global warming have been ridiculed, scorned, and berated as ignoramuses – or worse. The debate is "settled," announced Al Gore, acting as if under the illusion that he had been elected arbiter of all scientific disputes.[1] Those unwilling to get with the program – and on the gravy train – run the risk of becoming professional pariahs. Their treatment has been nothing short of shameful. The abuse of dissenters from this virtual religion achieved an apogee of sorts in 2007, when the columnist Ellen Goodman of the *Boston Globe*, noting that the apparently infallible United Nations had pronounced evidence for global warming as "unequivocal," wrote, "I would like to say we're at a point where global warming is impossible to deny. Let's just say that global warming deniers are now on a par with Holocaust deniers, though one denies the past and the other denies the present and future."[2]

The smug arrogance of Goodman and Gore – non-scientists who would cast into the outer darkness men and women of science who disagree with a theory whose policy implications accord with their own personal political agendas – is remarkable. It is obscene, even. It is also frightening. A dogma has been constructed from which the government–science establishment will brook no dissent – or even questioning. The columnist George Will notes, "Never in peacetime history has the government–media–academic complex been in such sustained propagandistic lockstep about any subject."[3]

But just as the high muckamucks of global warming were preparing for their greatest political triumph – the December 2009 United Nations Climate Change Conference in Copenhagen – a crack appeared in the mono-lith. In November 2009, a scandal broke that revealed just how politicized the climate debate had become. Emails from the Climate Research Unit (CRU) of the University of East Anglia in Norwich, the primary collector and disseminator of climate-change data, revealed a pattern of data-doctor-ing and dissembling in the service of the global warming agenda. Inevitably, the matter was dubbed "Climategate."

Some brave whistle-blower leaked approximately one thousand emails from the accounts of the Climate Research Unit. In his own words, Philip Jones, director of East Anglia's CRU, had employed statistical "tricks" upon data in order to mask evidence that current temperatures are *not* trending upward. At the very least, data had been massaged; at worst, some suspected, they had been falsified.[4] The anonymous leaker or leakers declared in a state-ment attached to the leaked emails, "We feel that climate science is too important to be kept under wraps. We hereby release a random selection of correspondence, code, and documents. Hopefully it will give some insight into the science and the people behind it."[5]

It sure did. As climate change skeptic Patrick Michaels said, "This is not a smoking gun; this is a mushroom cloud."[6] Among the more enlightening passages from the leaked emails was this remark by CRU climatologist Keith Briffa: "I know there is pressure to present a nice tidy story as regards 'appar-ent unprecedented warming in a thousand years or more in the proxy data' but in reality the situation is not quite so simple…. I believe that the recent warmth was probably matched about 1000 years ago."[7] This kind of candor, revealed to the world, was the CRU's worst nightmare realized.

This was no teapot tempest, no routine case of petty academic dishonesty to further a political agenda. The scientists involved were major presences in the field. "What we are looking at here," said Christopher Booker, writing in the *Sunday Telegraph* of London, "is the small group of scientists who have for years been more influential in driving the worldwide alarm over global warming than any others, not least through the role they play at the heart of the UN's Intergovernmental Panel on Climate Change (IPCC)."[8]

Philip Jones, director of the CRU, played a key role in the transmission of global temperature records to the IPCC, which, in its capacity as science advisor to the world's governments, has been a major force in pushing the global warming agenda. Incredibly, the original temperature data upon which the CRU has relied over the years – reports from weather stations – were destroyed in the 1980s, so scientists are forced to rely on second-hand information from what is now a compromised source. As University of Colorado environmental scientist Roger Pielke says, "The CRU is basically saying, 'Trust us.' So much for settling questions and resolving debates with science."[9]

Many leaked emails were innocuous or could plausibly be conceived as such, but others cast doubt upon the trustworthiness of the CRU researchers. One researcher complained, "The fact is that we can't account for the lack of warming at this moment and it is a travesty that we can't."

"Travesty" is a strange word to use in this case. *Travesty* means a grotesque imitation or parody. Unexplained developments in science are not "travesties." They are matters calling out for more research, more study, and careful evaluation. Whatever this is, it is not the language of honest science. It is, instead, as George Will writes, the language of political partisans who "consider it virtuous to embroider facts, exaggerate certitudes, suppress inconvenient data, and manipulate the peer-review process to suppress scholarly dissent and, above all, to declare that the debate is over."[10]

Calling this "the greatest scientific scandal of our age," Christopher Booker charged that global-warming scientists had intentionally manipulated data in order to strengthen their thesis – an act of shocking brazenness, but not, unfortunately, a complete surprise.[11] For consider the billions of dollars at stake. Leftist columnist Alexander Cockburn, who called the Copenhagen summit an "outlandish foray into intellectual fantasizing," wrote that "Billions in funding and research grants sluice into the big climate modeling enterprises. There's now a vast archipelago of research departments and 'institutes of climate change' across academia, with a huge vested interest in defending the [anthropogenic global warming] model. It's where the money is. Scepticism, particularly for a young climatologist or atmospheric physicist, can be a career breaker."[12]

Professor Ian Plimer, a geologist at Australia's University of Adelaide, charged that climate-change scientists were cooking data in order to essentially swindle taxpayers out of their money and enrich themselves. "The climate comrades are trying to keep the gravy train going," he told a London audience on the eve of the Copenhagen climate summit — which was conducted, as fate would have it, in a city beset by a blizzard. "Governments are keen on putting their hands as deep as possible into our pockets."[13]

Enormous sums — from a pot totaling perhaps hundreds of billions of dollars — are available to scientists who play their cards right, who go along with the consensus, who do not challenge received thought, who lobby and jockey for government favors with as much amoral nimbleness as any K Street lobbyist. The availability of such sums almost guarantees that unscrupulous persons will covet them, and that the worst of the bunch will lie, cheat, or steal to get their hands on the bounty. Others, perhaps savvier, will take a lesson from history and exaggerate threats to keep the pot of money expanding. The organizers of the Copenhagen summit — and of the Mexico City climate summit scheduled for 2010 (not in a blizzard, one hopes) — would have us believe that they were out to save the planet, to protect the rest of us from a future of melting icebergs and coastal floods and dead polar bears. The more immediate future, however, consists of conferences and publications and endowments and contracts and research that, because it is all subsidized by organizations that take a very definite view of the facts of global warming, will surely ratify those facts. The simple truth, known to all who seek a ride on the gravy train, is this: Dissenters do not receive funding. Period.

So what is to be done? As long as scientific hypotheses carry political implications, science will never be de-politicized. If it can be plausibly asserted that the temperature of the Earth is rising, and that this increase is at least partially attributable to, for instance, automobile emissions, then both the automobile industry and anti-automobile partisans will adduce favorable evidence from scientists — and, in all likelihood, some of that evidence will be the result of subsidized research. This is inevitable in the give and take of governance and scientific research. It can even be healthy, especially if the funding sources are acknowledged, as they will be by reputable researchers.

The involvement of government, however, complicates matters. Government funding of research, as we have seen, places the state on one side or the other of contentious matters. The coercive nature of government puts it in a different category from business, philanthropy, the academy, or private associations in the funding of research: it places the stamp of official approval upon certain doctrines (for instance, global warming) while effectively denying dissenters or skeptics the imprimatur of the state. Because central governments command vast sums of money, the disbursal of such monies will set off a mad scramble to secure such funds, and the resultant politicking and blacklisting and logrolling is antithetical to the supposed basis of the scientific enterprise: the search for truth. Government-subsidized science, no matter how pure-hearted and honorable some of its practitioners may be, is, in the end, the enemy of the free exchange of ideas.

The row over global warming is teaching this lesson not only to U.S. policymakers but to the world. In order to keep their tickets punched for rides on the gravy train, climate scientists are "massaging" data so that they do not contradict what has become the more or less official narrative of climate-change theory. The suppression of evidence in scientific debate is a serious matter. It calls into question not only the integrity of the scientists involved and their work, but also the system by which such research is funded. If government-subsidized science produces dishonest work — if its very incentive structure encourages dishonesty — then what is left of the case for such subsidy?

Big Science is so firmly entrenched at the government trough that dislodging it from its post may seem a truly quixotic notion. There are barely a handful, if even that, members of Congress who would endorse the views of the 19th-century critics of John Quincy Adams's "light-houses of the sky" and the Department of Science proposal of the 1880s. Contemporary Frank Jewetts are seldom heard in congressional hearings on science education or the space program or killer asteroids. That the federal government has a central role to play in such matters seems to be axiomatic.

But maybe it is time we dusted off the U.S. Constitution and listened once more to those who argued against entrusting the federal government with the direction of American science. Maybe we should take seriously the once-powerful case against creating national science bureaucracies. Maybe the idea of the scientist as government dependent is not as compelling as it seemed in the late 1940s, when Washington had just won a war and there appeared to be no limit to the capacities of the U.S. government. Maybe free men and women pursuing knowledge under private, corporate, philanthropic, and cooperative patronage is not quite the anachronism that it seemed to be in the Age of Big Government. Maybe it is time for another look at the virtues of private science. After all, that killer asteroid isn't due to arrive for another several million years, at least.

Notes

1. Jonathan Leake, "The Great Climate Change Scandal," *The Sunday Times* (London), November 29, 2009.
2. Ellen Goodman, "No Change in Political Climate," *Boston Globe*, February 9, 2007.
3. George Will, "The Climate-Change Travesty," *Washington Post*, December 6, 2009.
4. Leake, "The Great Climate Change Scandal."
5. "Climate Scientists Accused of 'Manipulating Global Warming Data,'" *Telegraph* (London), November 21, 2009.
6. Leake, "The Great Climate Change Scandal."
7. Quoted in Alexander Cockburn, "Turning Tricks, Cashing in on Fear," Counterpunch. com, December 18, 2009.
8. Christopher Booker, "The Worst Scientific Scandal of Our Generation," *The Sunday Telegraph* (London), Nov 29, 2009.
9. Jonathan Leake, "There's Just One Problem: CRU Dumped Climate Data," *The Sunday Times*, November 29, 2009.
10. Will, "The Climate-Change Travesty."
11. Booker, "The Worst Scientific Scandal of Our Generation."
12. Cockburn, "Turning Tricks, Cashing in on Fear."
13. John Ingham, "Climate Change 'Fraud,'" *Daily Express* (London), December 2, 2009.

Index

Breinigsville, PA USA
02 March 2011

256772BV00002B/86/P